Genetic Determinants of Human Longevity

Genetic Determinants of Human Longevity

Special Issue Editors

Serena Dato
Mette Sørensen
Giuseppina Rose

MDPI • Basel • Beijing • Wuhan • Barcelona • Belgrade

MDPI

Special Issue Editors

Serena Dato
University of Calabria
Italy

Mette Sørensen
Danish Aging Research Center
University of Southern Denmark
Denmark

Giuseppina Rose
University of Calabria
Italy

Editorial Office
MDPI
St. Alban-Anlage 66
4052 Basel, Switzerland

This is a reprint of articles from the Special Issue published online in the open access journal *Genes* (ISSN 2073-4425) in 2019 (available at: https://www.mdpi.com/journal/genes/special_issues/Genetic-Determinants-Human-Longevity).

For citation purposes, cite each article independently as indicated on the article page online and as indicated below:

LastName, A.A.; LastName, B.B.; LastName, C.C. Article Title. *Journal Name* **Year**, *Article Number*, Page Range.

ISBN 978-3-03921-678-9 (Pbk)
ISBN 978-3-03921-679-6 (PDF)

Contents

About the Special Issue Editors

Serena Dato obtained a Ph.D. in Molecular Bio-Pathology in 2004. Since September 2006, she has been an Assistant Professor in Genetics at the Department of Cell Biology of the University of Calabria, where she carries out research at the Genetics Laboratory. From the beginnning, her research interests have focused on the study of human longevity and in particular on the development of experimental designs and new analytical approaches for the study of the genetic component of longevity. With her group, she developed an algorithm for integrating demographic data into genetics, which enabled the application of a genetic-demographic analysis to cross-sectional samples. She was involved in several recruitment campaigns for the collection of data and DNA samples from old and oldest-old people in her region, both nonagenarian and centenarian families. She has several international collaborations with groups involved in her research field in Europe and the USA. Since 2008, she has been actively collaborating with the research group of Prof. K. Christensen at the Aging Research Center of the Institute of Epidemiology of Southern Denmark University, where she spent a year as a visiting researcher in 2008. Up to now, her work has led to forty-eight scientific papers in peer reviewed journals, two book chapters and presentations at scientific conferences.

Mette Sørensen has been active within ageing research since 2006, with work ranging from functional molecular biological studies to genetic epidemiology and bioinformatics. She obtained a Ph.D. in genetic epidemiology of human longevity in 2012 and was appointed Associate Professor at the University of Southern Denmark in March 2019. Her main research interest is in the mechanisms of ageing, age-related diseases and longevity, with an emphasis on genetic and epigenetic variation. Her work is characterized by a high degree of international collaboration and interdisciplinarity. The work has, per September 2019, led to thirty-one scientific papers in peer reviewed journal, as well as popular science communications, presentations at scientific conferences, media appearances, and an independent postdoctoral grant from the Danish Research Council in 2013.

Giuseppina Rose is Associate Professor in Genetics at the University of Calabria. She graduated from the University of Calabria School of Natural Science in 1983 and served as a Research Assistant there from 1992–1999. In 1994 she achieved a Ph.D. in Biochemistry and Molecular Biology at the University of Bari, Italy. She became Assistant Professor of Genetics at the University of Calabria from 2000–2003 before becoming Associate Professor in 2003. From the very beginning of her career, she worked on human population genetics, with a special reference in the genetic variability in complex traits. From 1996, she focused her interests in the study of human longevity and age-related traits, exploring the influence of mitochondrial and nuclear variability on human survival but also on the genetic predisposition to age-related complex diseases. Her studies have provided significant contributions to the identification of genetic factors modulating rate and quality of human ageing and have also contributed to recognize main players in molecular mechanisms related to ageing phenotypes. She has participated in several European projects, coordinating the local research activity and collaborating with group leaders in longevity studies. She has published eighty-eight scientific papers in peer reviewed international journals, four book chapters and more than one hundred contributions to scientific conferences.

Editorial

Untangling the Genetics of Human Longevity—A Challenging Quest

Serena Dato [1,*,†], Mette Soerensen [2,*,†] and Giuseppina Rose [1,*,†]

1   Department of Biology, Ecology and Earth Sciences, University of Calabria, 87036 Rende, Italy
2   The Danish Twin Registry and Epidemiology, Biodemography and Biostatistics, Department of Public Health, University of Southern Denmark, 5000 Odense-C, Denmark
*   Correspondence: serena.dato@unical.it (S.D.); msoerensen@health.sdu.dk (M.S.); pina.rose@unical.it (G.R.)
†   All authors equally contributed.

Received: 23 July 2019; Accepted: 25 July 2019; Published: 31 July 2019

Keywords: genetic determinants of human longevity; genetic variation; genetic association study; longevity-related genes; human lifespan

---

Human average life expectancy in developed countries has increased dramatically in the last century, a phenomenon which is potentially accompanied by a significant rise in multi-morbidity and frailty among older individuals. Nevertheless, some individuals appear someway resistant to causes of death, such as cancer and heart disease, compared with the rest of the population, and are able to reach very old ages in good clinical conditions, while others are not. Thus, during the last two decades we have witnessed an increase in the number of studies on biological and molecular factors associated with the variation in healthy aging and longevity. Several lines of evidence support the genetic basis of longevity: from the species-specific maximum lifespan to the genetically determined premature aging syndromes. Studies in human twins, that aimed to distinguish the genetic from the environmental component, highlighted a heritability of life span close to 25%. In centenarian's families, the offspring of long-lived individuals not only exhibit a survival advantage compared to their peers, but also have a lower incidence of age-related diseases. On the other hand, population studies found that genetic factors influence longevity in age- and sex-specific ways, with a most pronounced effect at advanced age and possibly in men compared to women. All this evidence indicates that a genetic influence on longevity exists, laying the foundation for the search for the genetic components of extreme long life. Consequently, over the past three decades, there has been a surge in genetic research, due in part to advances in molecular technologies, starting as studies of single genetic variants in candidate genes and pathways, moving on to array-based genome-wide association studies (GWAS) and subsequently to next generation sequencing (NGS). However, despite a plethora of studies, only few variants (in the *APOE*, *FOXO3A* and *5q33.3* loci) have been successfully replicated in different ethnic groups and the emerging picture is complex. For instance, it is an understatement to think that long-lived people harbor only favorable variants, completely avoiding risk alleles for major age-related diseases; indeed, there is evidence that many disease alleles are present in long-lived people. It is more probable that the longevity phenotype is the result of a particular combination of pro-longevity variants and risk alleles for pathologies, likely interacting in networks in a sex- and age-specific way. Finally, characteristics of aging are extremely heterogeneous, even among long-lived individuals, due to the complex interaction among genetic factors, environment, lifestyle, culture and resiliency. Population and study specificity, lack of statistical power for such a rather rare phenotype and missing heritability represent further hard obstacles to overcome in genotype–phenotype association studies. Thus, many challenges remain to be addressed in the search for the genetic components of human longevity. In this Special Issue we included five original articles and two reviews covering different areas in the field of human longevity, to help the reader take stock of the situation and point to future perspectives of the field.

The two reviews in particular look closely at two main arguments of biogerontology research. The paper by Taormina and co-workers gives an updated review of the lessons from model organisms, where a substantial number of findings suggest that longevity could "directly" be promoted by interventions in specific pathways, like inflammation, oxidative stress response, DNA repair, as well as the use of nutrients [1]. The most relevant model systems used in biogerontology are discussed, as well as significant discoveries confirmed in humans, advising the researchers to use different model systems to avoid misinterpretation of the results due to confounding factors or model system peculiarities. The paper by Abondio et al. [2] reviews the available literature on *APOE* and its involvement in biological pathways related to human longevity, under an anthropological and population genetics perspective, highlighting the evolutionary dynamics, which may have shaped the distribution of its haplotypes across the globe and the potential adaptive role. Both of the two reviews are useful compendia of reference papers in the field and, at the same time, provide good points of discussion for future studies on the genetic aspects of human longevity.

The paper by Hjelmborg et al. discusses an interesting topic for demographical studies on human longevity, i.e., the role of zygosity in a twin's lifespan [3]. Twin cohorts have been analyzed in several GWAS of common traits and diseases around the world. Although there is no evidence that the gene–disease associations seen in singletons differ in twins, the question if selecting one individual from a twin pair implies a selection in survival due to zygosity is still often questioned. The authors compared the relative survival of monozygotic (MZ) with dizygotic (DZ) twins (from the 1870–1900 and the 1961–1990 birth cohorts), from one of the largest nationwide cohorts of twins with valid vital status, the Danish Twin Registry; they found no correlation of mortality with zygosity, meaning that MZ and DZ pairs appear to share the same mortality process. Thus, being a twin does not appear to impact the basic biological processes and human development in adolescence and adulthood. This is an interesting result for the studies on disease onset and other age-dependent traits which use twins, because it implies that findings from twins are generalizable to the population as a whole, especially when large sample sizes are used.

The three candidate gene association studies of longevity presented here (the papers by Scarabino et al., De Rango et al. and Crocco et al.) directly deal with the search for the genetic component of longevity and healthy aging. The response to external injuries is the leitmotif unifying all three association studies.

As it is known, internal and external stresses disrupt telomere homeostasis. The contribution by Scarabino and co-workers confirms the genetic determination of leukocyte telomere length (LTL) by *TERT* variability, showing that shorter LTL at baseline may predict a shorter lifespan [4]. Furthermore, they found that the reliability of LTL as a lifespan biomarker could be age-specific and act in specific age-spans (age 70–79 in their study population).

The papers by De Rango et al. and Crocco et al. highlight the complexity of longevity, a highly dynamic phenotype influenced by internal and external stresses, that makes the identification of genes robustly associated with it very challenging. De Rango's paper investigated for the first time the contribution to the longevity phenotype of the genetic variability of *IPMK* (Inositol Polyphosphate Multikinase), a potential moonlighting protein performing multiple functions in pathways affecting the aging process, from nutrient-sensing to oxidative stress and telomere maintenance [5]. This paper supports this gene as a novel gender-specific determinant of human longevity on one hand, and on the other hand promotes pleiotropic proteins like IPMK, able to integrate cellular activities in space and time, as crucial determinants of the complex connections among aging, health, and longevity. Dynamic genetic effects on longevity were found in the paper by Crocco et al. who investigated the variability of xenobiotic metabolizing genes, known to mediate the response/toxicity to xenobiotics [6]. They found lifelong changes in the frequency of alleles at *CYP2B6*, *CYP3A5*, *COMT* and *ABCC2* genes, following either linear or non-linear trajectories with respect to the chance of becoming long-lived. Such findings underline once again that SNPs associated with longevity might behave either as pro-longevity or killing variants but also as deleterious variants neutralized by the protective effect of pro-longevity genes (buffered variant), an important aspect to take into account in disentangling the genetic contributors to human longevity.

Finally, the paper by Revelas and co-workers contributes to the open debate of whether extreme longevity is coupled to risks for major diseases [7]. The authors built a polygenic risk score for cardiovascular health, based on GWAS variants, cardiovascular-related risk factors (such as cholesterol levels) and cardiovascular multi-morbidity disease (myocardial infarction, coronary artery disease, stroke etc.). By exploring the genetic profiles of 95+ year old individuals, the authors found that these extremely long-lived subjects did not have lower polygenic risk for cardiovascular health as compared to younger subjects, thus supporting the theory that exceptional longevity does not necessarily imply the absence of risk factors for major age-related traits.

Overall, we believe these papers, highlighting different facets and the complexity of the studies on the genetics of human longevity, may help to understand the path research has taken in the field up to now, and to explore some possible interventions, taking advantage of the pathways highlighted and of the perspectives they are unveiling for the future.

So, where should the longevity genetic field go from here? Despite decades of genetic studies, the variants consistently identified as associated to human longevity explain only a small part of the estimated heritability. For facing the challenge of 'missing heritability', new and innovative approaches are needed. First of all, more studies are needed to elucidate the effect of rare variants with larger effect sizes, not captured by standard GWAS. The opportunity of high-throughput methodologies like NGS, together with large multi-center collaborative studies of extremely long-lived cohorts, can contribute to pave the way for untangling the networks involved in human longevity. In this sense, the study of epistatic effects of different genetic markers in gene-set and pathway-enrichment analyses, as well as the integration of several layers of biological variation (SNP, Copy Number Variants, epigenetic markers) in polygenic risk scores could further help too. The application of a more functional genomic approach like the collection of whole-exome sequencing, genome-wide epigenetic, cell-specific transcriptomic data and the integration of all these layers of genomic information can help to disentangle the determinants of lifespan. Finally, the collection of life-long environmental and lifestyle variables known to influence an individual's health (like microbiome and nutrition), can significantly improve our chance to untangle the intricate interplay between genes and environment in determining the longevity phenotype.

Funding: This research was funded by the Italian Ministry of University and Research (PRIN: Progetti di Ricerca di rilevante Interesse Nazionale—2015, Prot. 20157ATSLF) to G.R. M.S. was supported by The Danish Council for Independent Research—Medical Sciences (DFF-6110-00016).

Conflicts of Interest: The authors declare no conflict of interest.

References

1. Taormina, G.; Ferrante, F.; Vieni, S.; Grassi, N.; Russo, A.; Mirisola, M.G. Longevity: Lesson from Model Organisms. *Genes* **2019**, *10*, 518. [CrossRef] [PubMed]
2. Abondio, P.; Sazzini, M.; Garagnani, P.; Boattini, A.; Monti, D.; Franceschi, C.; Luiselli, D.; Giuliani, C. The Genetic Variability of *APOE* in Different Human Populations and Its Implications for Longevity. *Genes* **2019**, *10*, 222. [CrossRef] [PubMed]
3. Hjelmborg, J.; Larsen, P.; Kaprio, J.; McGue, M.; Scheike, T.; Hougaard, P.; Christensen, K. Lifespans of Twins: Does Zygosity Matter? *Genes* **2019**, *10*, 166. [CrossRef] [PubMed]
4. Scarabino, D.; Peconi, M.; Pelliccia, F.; Corbo, R.M. Analysis of the Association Between *TERC* and *TERT* Genetic Variation and Leukocyte Telomere Length and Human Lifespan-A Follow-Up Study. *Genes* **2019**, *10*, 82. [CrossRef] [PubMed]
5. De Rango, F.; Crocco, P.; Iannone, F.; Saiardi, A.; Passarino, G.; Dato, S.; Rose, G. Inositol Polyphosphate Multikinase (*IPMK*), a Gene Coding for a Potential Moonlighting Protein, Contributes to Human Female Longevity. *Genes* **2019**, *10*, 125. [CrossRef]

6. Crocco, P.; Montesanto, A.; Dato, S.; Geracitano, S.; Iannone, F.; Passarino, G.; Rose, G. Inter-Individual Variability in Xenobiotic-Metabolizing Enzymes: Implications for Human Aging and Longevity. *Genes* **2019**, *10*, 403. [CrossRef]

7. Revelas, M.; Thalamuthu, A.; Oldmeadow, C.; Evans, T.J.; Armstrong, N.J.; Riveros, C.; Kwok, J.B.; Schofield, P.R.; Brodaty, H.; Scott, R.J.; et al. Exceptional Longevity and Polygenic Risk for Cardiovascular Health. *Genes* **2019**, *10*, 227. [CrossRef] [PubMed]

GCAT
TACG
GCAT
genes

MDPI

Review
Longevity: Lesson from Model Organisms

Giusi Taormina, Federica Ferrante, Salvatore Vieni, Nello Grassi, Antonio Russo and
Mario G. Mirisola *

Dipartimento di Discipline Chirurgiche, Oncologiche e Stomatologiche, Università di Palermo,
Via del Vespro 129, 90100 Palermo, Italy
* Correspondence: mario.mirisola@unipa.it

Received: 10 April 2019; Accepted: 2 July 2019; Published: 9 July 2019

Abstract: Research on longevity and healthy aging promises to increase our lifespan and decrease the burden of degenerative diseases with important social and economic effects. Many aging theories have been proposed, and important aging pathways have been discovered. Model organisms have had a crucial role in this process because of their short lifespan, cheap maintenance, and manipulation possibilities. Yeasts, worms, fruit flies, or mammalian models such as mice, monkeys, and recently, dogs, have helped shed light on aging processes. Genes and molecular mechanisms that were found to be critical in simple eukaryotic cells and species have been confirmed in humans mainly by the functional analysis of mammalian orthologues. Here, we review conserved aging mechanisms discovered in different model systems that are implicated in human longevity as well and that could be the target of anti-aging interventions in human.

Keywords: model systems; aging; signal transduction; molecular senescence

1. Introduction

Aging is considered a natural and unavoidable "side effect" of life in spite of the observation that life span can vary greatly between species and individuals. Researchers have developed many theories on the cause of aging, but none of them prevailed. George Williams, an undisputed leader of the field, proposed that natural selection fixes alleles in a population for their positive contribution to fitness early in life, but that the selected alleles become deleterious later in life [1]. Around this idea, he developed the antagonistic pleiotropy theory. According to this proposal, aging is the result of (a) casual selection of late-acting deleterious alleles because of their advantage early in life and (b) the incapacity of natural selection to filter the late-acting detrimental effect of these alleles. Corollary to this theory is that mutations that are capable of increasing the life span must cause a disadvantage or a reduced fitness during the early stages of life. Many mutations that are capable of increasing the life span of different model systems confirmed this prediction, but the existence of long-lived mutants, such as the yeast Ras2 or the *C. elegans* daf-2 that grow and reproduce at a normal rate [2], suggests that extended life doesn't imply a reproductive or growth rate fitness cost. However, these observation are made in laboratory conditions; therefore, we cannot exclude that these mutants could also show a reduced fitness, but only in the wild.

The disposable soma theory proposes aging as a stochastic process driven by the progressive accumulation of molecular damages within somatic cell lines. This damaging force can be counterbalanced by repair systems that tend to maintain the cellular status quo at some energetic cost. Therefore, the aging rate, which profoundly differs between species, is inversely associated with the amount of energy used for somatic maintenance. According to this theory, periods of nutrients shortage forced organisms to balance the energy used for germ line maintenance as well as reproduction and somatic maintenance [3]. Programmed longevity theory has been proposed as an alternative [4]. It proposes that the healthy portion of life span is programmed to increase the fitness. The aging rate wouldn't be regulated by an energy-based trade-off between reproduction and longevity, but each species reaches an evolutionary stable strategy. According to this theory, most longevity-extending mutations will cause a trade-off, but some won't.

However, there is the possibility that the group instead of the individual is the object of evolution. According to this theory, fitness is affected by either "group competition", "individual competition", or both, depending on the conditions encountered. The advantage of the single individual is positively selected only if it also confers a group advantage. Mathematical simulations and experimental evidences suggest that at least in *S. cerevisiae*, aging under certain conditions can be programmed, and an altruistic life span extension of the individual can provoke the extinction of the group [4].

None of these theories prevailed so far, which was probably because the complex forces of evolution cannot be simplified by a single scheme. Evolution may be the result of multi-level selection where individual fitness or group fitness are preferred, depending on ecological niche and population density.

Model systems have been widely used to confirm these theories. Their advantage relies on reduced costs, easy maintenance in a laboratory facility, and affordable but rigorous manipulation tools. In general, the simpler the organism we use as a model, the greater these advantages. The short life span of simple organisms is another key advantage of model systems in aging research (Figure 1), which allows both parallel and serial experiments to be performed in reasonable time. In addition, rigorous genetic experiments can be performed in simple eukaryotes using the powerful possibilities of modern molecular genetics, which permit simultaneously screening multiple congenic, or even in some models such as yeasts or isogenic individuals at relatively low cost. The other side of the scale is that some of the simpler systems cannot model alone all the aging phenotypes (e.g., immunosenescence can't be modeled by yeasts). It is not surprising that, as in many other science fields, biogerontologists thus widely take advantage from the usage of multiple model organisms to explore their theories. In addition, ethical issues are of course much more limited in simple model organisms making *Saccharomyces cerevisiae*, *Caenorhabditis elegans*, *Drosophila melanogaster*, and *Mus musculus* the most popular model systems used in aging research. *Canis lupus familiaris* is becoming an interesting alternative to those model systems because, even though it has a similar ethical constraint to that of humans, massive breed selection offer interesting genetic opportunities. We will describe below the genes and pathways discovered thanks to these organisms that have shed light on possible aging mechanisms in humans.

2. The Simplest Eukaryotic Model: Yeast Cells

Yeast has played a critical role in revealing the molecular genetics of many basic cellular processes such as the cell cycle [5], protein folding [6], intracellular trafficking [7,8], and many others. A short generation time, easy and cheap lab setups, powerful genetic approaches, and high-throughput methodologies [9] are some of the attractive factors of this simple unicellular eukaryote. Genome similarities (homologue and orthologue genes) between this very simple organism and mammalians or even humans are surprisingly high, suggesting that this organism can be an effective model of human diseases [10–12]. Two different paradigms of aging exist in yeast: replicative life span (RLS) and chronological life span (CLS) [13]. The first one measures the replicative potential of individual cells. Single cells can in fact be monitored for their ability to generate new cells. Almost 60 years ago, Robert K. Mortimer used this characteristic to count the number of cell divisions obtained from a single cell [14]. In this protocol, the emerging bud (daughter cell) is carefully removed by micromanipulation as it appears until the mother cell stops dividing. The second paradigm (CLS) measures instead the survival of a cellular population in the post-mitotic, non-dividing phase [15]. Thus, the yeast RLS resembles the Hayflick limit observed in cultured mammalian cells, measuring essentially the doubling ability of cultured cells, while the CLS models better the ability to survive and retain the doubling potential of post-mitotic tissues. It is interesting to note that in spite of the clear differences of the two paradigms used to summarize aging in yeast, some pathways such as Ras-PKA or Tor-Sch9 have a consistent role in CLS and RLS [13,16], while others do not (e.g., Sir2 [17]). The yeast life span is doubled through deletion of the gene encoding the small G-protein

Ras2, showing a contemporary higher resistance to multiple stresses [18,19]. Ras proteins are molecular switches that can cycle between the on–off state, which correspond to the GTP-bound and GDP-bound, respectively [20]. Mutations on the human orthologue of the yeast Ras are found in 35% of human cancers. The amount of the Ras on-state molecules is dependent on the availability of sugars [21]. Two different pathways are known to be dependent on Ras2 activation in yeast: protein kinase A (PKA) and Mitogen-activated protein kinase (MAPK). Many data suggest a major role of the PKA pathway as a longevity regulator [22]. Msn2/Msn4, two key stress resistance transcription factors, are in fact inhibited by kinase A activity, and Msn2/4 deletion reverts the phenotype observed with Ras2 deletion. In addition, mutations affecting adenylate cyclase, an activator of the kinase A pathway, also increased the survival of yeast cells, confirming the role of the PKA pathway as a major Ras-dependent pathway relevant to the aging processes. PKA activation leads to Rim15 inhibition which, in turn, positively regulates the transcription factors Gis1 [23], Msn2, and Msn4, whose function is to bind respectively the PDS (post-diauxic shift) and STRE (STress Responsive Element) sequences activating the expression of genes involved in survival and stress response such as heat shock proteins, the cytoplasmic catalase, and the two superoxide dismutase (SOD1, SOD2) [24,25], or in DNA repair such as the DDR2 gene [26].

A selection of stress-resistant mutants obtained by transposon mutagenesis was evaluated by chronological life span to isolate long-lived mutants. In this study, the Sch9-Tor pathway emerged as another pro-aging pathway [27,28]. Sch9 is a serine/threonine protein kinase that is homologous to mammalian Akt and S6K. It has been isolated in yeast as a multicopy suppressor of Ras/PKA through the conditional inactivation of the thermosensitive Ras exchange factor allele CDC25ts [29,30]. In addition, the increased G1 length of yeast cells bearing a Sch9 null allele is suppressed by a hyperactive kinase A pathway. These and other data suggest that its function is partly overlapping with the Ras/PKA pathway. The down-regulation of both signaling pathways obtained by specific nutrients depletion results in the greater expression of genes involved in resistance to stresses [31]. The observation that SOD2 expression levels doubles in ras2Δ strains suggests that oxidative stress is a major component of this response [32]. In addition, overactivation of the transcription factors that respond to thermal (HSF1) and oxidative stresses (YAP1) increases survival [33,34]. In spite of these observations, the over-expression of superoxide dismutases and catalases alone does not allow the survival levels obtained by the deletion of RAS2 and SCH9, suggesting that modulation of oxidative stress is not the only mechanism involved in longevity regulation. In addition, despite the aforementioned functional overlap between the Ras and Sch9 pathways, the simultaneous deletion of RAS2 and SCH9 has greater effect than their respective individual deletions. It is important to notice once again that the mediators that are part of these pro-aging cell pathways find their functional or structural orthologues in higher eukaryotes, and it has been confirmed that the mechanisms described are conserved from yeast to mammals. In addition, it has been observed that in all organisms, these pathways are influenced by nutrients, either directly or through insulin/insulin-like growth factor-1 (IGF-1) and growth hormone (GH) in multicellular eukaryotes [35–37].

3. Caenorabtidis elegans

Caenorhabditis elegans has been established as a model organism in 1965 by Sidney Brenner. This small size (1.5-mm long adult) soil nematode has great potential for genetic analysis, is easy and cheap at laboratory cultivation, and has a short life cycle: the generation time is only three to five days, and the life span is two to three weeks. *C. elegans* can inbreed by self-fertilization (hermaphrodite). Alternatively, hermaphrodites can cross with males (a possibility that is otherwise known only in plant genetics where selfing and crossing can be manipulated at will). Other key features are the anatomical simplicity (<1000 cells) and the small genome. The latter is about 101-Mbp long, consists of six chromosomes, and contains 19,000 genes, of which 50% are conserved in the human genome. The animal body is transparent, which is a characteristic that allows the tracking of cells in vivo over time, and the fluorescent visualization of tagged proteins. It can be cultivated on agar plates or in liquid media; therefore it is a good system for a wide variety of high-throughput manipulations.

In addition, the researchers take advantage of its powerful and continuously expanding genetic and imaging possibilities [38,39]. First, a multicellular organism is completely sequenced, revealing a conservation of ~80% of its proteins with vertebrates [40]. RNA interference (RNAi) was the most commonly used approach to address gene function; however, tissue-specific knockdowns have recently been developed [41]. Another reason for its wide usage as a model system is that simple phenotypes can be observed, easing the analysis of most genetic screens.

Many key factors that are capable of regulating longevity have been discovered using this model organism [42,43]. In 1993, it was demonstrated that mutations at the daf-2 locus, which encodes an ortholog of the insulin/IGF1 receptor, doubles the life span of the animal [44–46]. Further molecular characterization revealed that the life-span extension observed with defective daf-2 alleles required the activity of the daf-16 gene product [44]. The latter is a transcription factor of the FOXO (forkhead box transcription factor O) family, giving the first example of transcriptionally regulated aging modulation [47]. The FOXO family is an evolutionary conserved group of transcription factors that target the protein kinase Akt. In the presence of growth stimuli, FOXO proteins are first exported from the nucleus to the cytoplasm, and then degraded via the ubiquitin–proteasome pathway. On the contrary, in the absence of growth stimuli, FOXO proteins are imported to the nucleus, where they up-regulate a group of target genes, ultimately promoting cell cycle arrest, stress resistance, or apoptosis. Stress stimuli also trigger the relocalization of FOXO factors into the nucleus, thus allowing an adaptive response to stress stimuli. Additional studies identified the PI-3 kinase signaling pathway as the downstream molecular cascade target of the DAF-2 receptor, provoking the nuclear localization of the mentioned DAF-16 transcription factor [48–50]. HSF1, the *C. elegans* orthologue of the heat shock transcription factor, plays a critical role in the DAF-2 observed regulation of longevity. It has also been demonstrated that HSF-1 is capable of increasing the life span of DAF-2 pathway mutants, suggesting that heat shock proteins may have a pleiotropic role on pathways other than the daf-2 molecular cascade [51]. At the same time, mutations in daf-18, which is a homolog of PTEN (Phosphatase and tensin homolog) phosphatase that normally negatively regulates the insulin-dependent signal, suppress the increased survival due to mutations on DAF-2 and age-1 loci [52,53]. The latter codes the *C. elegans* catalytic subunit of the phosphatidylinositol-3-OH kinase (PI(3)K), whose impairment triggers constitutive Dauer status (the larva goes into a type of stasis and can survive harsh conditions), and increases life span and stress resistance. The lengthening of the survival of these mutants has been related to their increased resistance to oxidative stress; in fact, DAF-2 mutants express high levels of antioxidant enzymes such as catalase and superoxide dismutase, as well as low levels of free radicals. Mutants at the age-1 locus prevent the age-related decrease of catalase levels [54]. Other proteins that have been identified as important for longevity in nematode include the protein kinase phosphatidylinositol-dependent 1 (PDK1), the kinase Akt, and the mitochondrial enzymes Clk involved in ubichinone synthesis. In particular, the reduction in Clk-1 function produces smaller worms that live 15–30% longer than wild types, which is probably thanks to a decrease in basal metabolism and oxidative damage [55].

4. Drosophila melanogaster

Drosophila melanogaster has a 100 year-long history as an important model organism for studies on genetics and molecular biology. In 1908, Thomas Hunt Morgan was the first to use *Drosophila* as a model organism, demonstrating that genes are located on chromosomes. The fruit fly has also been the first organism for which a genetic map was obtained, thanks to Alfred Sturtevant and Calvin B. Bridges. *Drosophila melanogaster* has a generation time of 10–14 days and shows a high rate of reproduction. Large numbers of flies can easily be cultivated. Maintenance is quite simple and cost-effective. *Drosophila*'s genome is very simple: four pairs of chromosomes, 13,000 genes, and about 170 Mbp; it is about 20-fold smaller than a typical mammalian genome, but in spite of that, it encodes approximately the same number of gene families, thus making it easier to study gene function. Furthermore, many pathways, tissues, and organ systems in *Drosophila* are shared by mammals; approximately 60% of the genes that are known to be involved in human diseases have functional orthologues in the fruitfly. Thus, *Drosophila* represents one of the most useful and popular model organisms for research on human diseases. Aging research further benefits from its short life span (4–6 weeks), which allows manipulating and observing several generations while monitoring the effect of drugs, nutrients, genetic manipulation, as well as environmental factors over time. Other advantages are the wide range of tools to modulate gene function (such as mutagenesis screens, RNA interference, and transgene expression in specific stages of life or in selected tissues) and resources (cell lines, clone libraries, antibodies, microarrays, and databases) [56,57].

Studies on *Drosophila* show that longevity increases as a result of the over-expression of genes involved in stress response such as hsp70 (which encodes the heat shock protein 70), MnSOD (producing the superoxide dismutase), and mei-41 (involved in DNA repair) [58]; meanwhile, the loss of function of the pathway that includes the hormone receptor Dts3 and the insulin receptor (InR), Chico (which is the substrate of InR [59,60]), and the transcription factor dFOXO, greatly increases the life span [61]. Giannakou and Hwangbo in 2004 demonstrated that the tissue-specific overexpression of FOXO in gut and fat extends the life span, which is likely through the transcriptional repressor aop [62], which shares some targets with FOXO. Among these, Obp99b seems to have a key role, since it is upregulated in a long-lived *Drosophila* model [63]. mTOR inhibition also by rapamycin administration [64] extends life spans [65], but only in the presence of the s6 kinase [65,66]. The life extension by rapamycin treatment also requires an intact autophagy pathway [64,67]. Dietary restriction can increase the longevity of *D. melanogaster* by up to 30% as well as reduce the reproduction rate [68]. The restricted intake of specific nutrients, particularly proteins, may mediate the benefits as well as alter the macronutrients ratio. Amino acids and some amino acids sensors, such as Gcn2, seem to have a key role in life-span extension by dietary restriction [69].

5. Mouse

Domestic mouse is the most widely used mammalian model. Its genome is almost the same size as the human one (2.5 Gbp and 40 chromosomes), and encodes essentially the same number of genes. Most (85%) of protein-coding regions of the mouse genome are identical to human genome; in addition, 99% of the 25,000 genes have a human orthologue. Compared to other model organisms, working with mice is more difficult in every respect. They are bigger, have a generation time of about 8–10 weeks, and produce, on average, only six to eight young per brood. These numbers are interesting when compared to other mammals, but aren't when compared to simpler model organisms. Colonies of mice are expensive to maintain, and their genetic manipulation is quite difficult. However, unlike the fruit flies and nematodes, mice have an immune system, musculoskeletal apparatus, endocrine system, digestive system, and even nervous system similar to humans both in function and architecture. A fundamental tool to explore the genetics of aging in mammals is the identification of single-gene mutations that increase life span. The first of these was discovered in 1996 studying the Ames dwarf mouse [70]. Ames dwarf mice produce a reduced amount of growth hormone (GH), prolactin, and thyroid-stimulating hormone; they have a recessive point mutation in the Prop1 gene leading to hereditary dwarfism [71]. Prop1 encodes a protein required for the pituitary activation of Pou1f1, which is a member of the POU family of transcription factors implicated in mammalian development [72]. In 1996, Brown-Borg et al., following the life span of 62 Ames dwarf mice, showed a mean life span increased of 49% and 68% in males and females respectively, and a maximal life span extension of 20% and 50%, respectively [70]. This was the first demonstration that a single gene mutation can extend the life span in mammals. The Snell dwarf mouse was described in 1929; it has a recessive spontaneous point mutation in the Pou1f1 gene that causes hereditary dwarfism [73]. Flurkey et al. in 2001 followed the life span of 24 Snell dwarf mice together with 33 normal controls. He discovered that the mean life span was extended in the dwarf group by 48%, and some of the common age-related declines were delayed [74,75]. Other studies highlight that the dwarf mice have alterations in insulin/IGF-1 signaling: both Ames and Snell dwarf mice have severely reduced circulating levels of insulin, IGF-1, and glucose [76,77]. In addition, the dwarf mice show a slower metabolism [78]. Most of the dwarf mice are also infertile [79], but they display a lower spontaneous mutation frequency [80] and a very low incidence of malignant lesions compared to controls at the time of death [81]. Notably, dwarf mice have reduced levels of DNA and protein oxidation in the liver compared to control mice [82], according to increased levels or activity of catalase and Cu/Zn superoxide dismutase in various tissues at different ages [83,84], which makes these mice resistant also to the effects of chemical stressors [85], and shows low levels of reactive oxygen species. However, the importance of oxygen concentration as a driver of protein and DNA damages leading to cellular senescence has recently been challenged [86]. In fact, it has been demonstrated that while lab mice-derived primary fibroblasts are sensitive to oxygen concentration, primary fibroblasts from wild-caught mice are not. In addition, cells from other wild-caught rodents have lower oxygen sensitivity than lab mice-derived cells [86]. This observation underlines the need to use multiple model systems and the necessity, when possible, to confirm the results on wild-caught animals, since the repeated breeding and the laboratory condition of lab animals may trigger unwanted genetic drift.

An example of the results confirmed in multiple systems are the metabolic pathways involved in the aging of *Drosophila* and *C. elegans*, which were confirmed in mice thanks to genetic manipulation. Among the genes whose deletion significantly increases the survival of mice, some of the genes involved in stress resistance have been identified, such as for example GPx4, which codes for glutathione peroxidase, and some that are part of the insulin pathway, such as FIRKO (insulin receptor in adipose tissue) [87]. Animals that have mutations in the growth hormone receptor also live longer and show a lower incidence of age-related cognitive impairment and improved insulin sensitivity [88], while those who over-express GH show signs of accelerated aging [77]. Also, female Irs-1 knock-out mice (Irs-1 is the major intracellular effector of insulin) live 32% longer than the wild-type counterpart. Finally, inhibition of the mTOR (mammalian target of rapamycin) pathway, which was obtained by rapamycin administration through S6 kinase deletion, increases survival and reduces the incidence of age-related diseases, including immune dysfunction and insulin resistance [89]. Inactivation of the PKA pathway also increases survival and causes a reduction of tumor incidence and insulin resistance over time. Another key aging gene in mice is klotho. This protein is a circulating factor that inhibits the intracellular insulin/IGF-1 signaling cascade, but its function is still controversial. Homozygous klotho-deficient mice display a syndrome similar to human progeria and anomalies in different tissues. These mice are short-lived and infertile; they show growth retardation, premature thyme involution, ectopic calcification, skin atrophy, arteriosclerosis, osteoporosis, and pulmonary emphysema [90]. On the other hand, klotho gene over-expression extends mice life span by 20–30% [91]. Higher levels of klotho are associated with longer life span, reduced atherosclerosis risk, and better hearing than other mouse strains [92]. Some mutants with mutations in DNA maintenance, stability, and repair exhibit premature aging phenotypes. Wrn knock-out mice lightly display characteristics of premature aging, including the contemporaneous deletion of the Terc gene, which encodes the RNA component of the telomerase enzyme, and brings Werner's phenotype [93].

6. Domestic Dog

The domestic dog is a very interesting aging model for different reasons. Dogs allow us to observe them in their natural environment while being investigated; notably, they often share lifestyle and sometimes exercise habits with humans. Canine life span can vary greatly [94]. Interestingly, it has been observed that life span inversely correlates with body size. Larger breeds live almost six to seven years, while smaller breeds can reach up to 16 years [95–97]. Dogs share the same diseases with humans, in particular age-related diseases such as congestive heart failure, renal and hepatic disease, sarcopenia, diabetes, obesity, joint disease, neurodegeneration, cataracts, immune-mediated illnesses, and cancer [97–101]. In addition, it is the only species where massive breed selection has led to large numbers of individuals with very small phenotypic and genotypic differences within a specific breed, but with very large differences between different breeds. Maybe as a result of this strong selection for specific traits, some pure breeds of dogs are more prone to specific age-related diseases than others [97]. For example, the analysis of cancer incidence in different pure breeds confirmed that larger breeds have a higher incidence of cancer but revealed also that each breed has increased incidence for specific cancer types, suggesting that anatomic differences and genomic differences between breeds can explain these differences. It is also interesting to know that as a companion animal, dogs often share part of their lifestyle with their owners, including physical exercise. At the same time, dogs receive medical treatments from veterinarian specialists during their life as it happens for humans, and that contributes to increase the life span of these companion animals. The domestic dog has a well-annotated genome that was fully sequenced 10 years ago [102–104]. The size of the diploid genome is 4900 Mbp, which is organized in 38 pairs of autosomes and two sex chromosomes. Different genetic resources are applied in canine genetic research. Furthermore, genetic pedigrees are registered for many generations [105]. Greer et al. (2007) reported a correlation between body size and longevity. They noticed that smaller members (Chihuahua) live longer than larger members (e.g., Great Dane) [95]. Similar data have been obtained from studies in a murine system in which the IGF/GH axis resulted involved in longevity and body size determination [106]. Notably, circulating IGF-1 levels correlate with the body size of adult dogs; in fact, IGF-1 levels increase with body weight. Therefore, the data is consistent with previous work by Eigenmann et al. (1988), indicating that adult dogs have a high correlation between circulating IGF-1 and adult body size [107]. Furthermore, IGF-1 decreases over time as a function of age [108]. In addition, a reduction of the insulin/insulin-like growth factor-1 (IGF-1) signaling cascade significantly extends life span [109]. Interestingly, Greer et al. described that serum IGF-1 levels decrease at a higher rate in intact females than in spayed females. Similarly, there is a significant difference in the serum IGF-1 levels between neutered and intact males. For both intact males and females, an increase in overall body weight was significantly associated with higher levels of IGF-1. These data suggest relevant hormonal effects on IGF-1 action [104]. Interestingly, Waters et al. (2009) demonstrated that dogs with four years or more of ovarian hormone exposure live longer than ovariectomized dogs [110]. Therefore, higher body weight is related to high level of serum circulating IGF-1, which in turn seems to be deleterious for aging. Accordingly, the maintenance of lean body mass and reduced accumulation of body fat have been associated with attaining a longer than average life span [111]; thus, dog's weight may be more predictive of life span than height, as reported by Greer et al. [95]. A dietary restriction of 25% has been reported to increase life span by about two years [112,113], improving metabolic health [114,115] and delaying immune senescence in Labrador retrievers maintained in a laboratory environment [114,115]. Further studies will be necessary to discriminate as to whether increasing the health and life span after CR in dogs is the effect of weight loss or depends on the down-regulation of pro-aging pathways such as previously demonstrated in simple model organisms such as yeast. Actually, studies carried out by Jimenez et al. suggest that large breed dogs may have higher glycolytic rates, and an increased DNA mutation rate, which could be responsible for their decreased life span compared with small breed dogs, despite reactive oxygen species (ROS) production showing no differences across size and age classes [116]. Likewise, Alexander et al. observed a decline in the heat shock protein 70 response after heat stress with age, suggesting a role of oxidative stress in dog's aging [117].

7. Non-Human Primates

The models closest to humans are non-human primates. In fact, the chimpanzee genome has more than 98% homology with the human genome, while the rhesus monkey (*Macaca mulatta*) has 93% DNA homology with the human genome [118,119]. The Macaca mulatta genome is 3146 Mbp and codes 21,000 different proteins [120]; *Pan troglodytes*, a common chimpanzee, has a 3385-Mbp genome coding for 23,534 proteins. Interestingly, they have a similar inter-individual variability to humans and share eating and sleeping behavior, physiology, neurology, endocrinology, immunology, as well as anatomy [121]. Their average life span ranges from 7 to 30 years depending on the species considered, while the maximum life span of the commonly used macaque reaches 40 years, and the chimpanzees can reach up to 65 years [122].

Similarly to humans, monkeys exhibit signs of physical decline and many age-associated diseases, including [123]: sarcopenia [124], osteoporosis [125], cataracts, cardiovascular diseases, and cancer [123]. In fact, the incidence of these diseases is similar to that in humans. Researchers take advantage of two important resources: the Internet Primate Aging Database, which collects data such as body weight and blood chemistry measurements on many non-human primate species during aging, and the Nonhuman Primate Tissue Bank. Another advantage is the possibility to fully control the environment, dietary intake, and medical history. On the other hand, non-human primates require specialized care, have high cost, and require ethical considerations. However, they represent the ideal model to summarize the complex in vivo physiology of human aging. Even though research such as direct manipulation of the monkey genome are not possible for ethical reasons, observational studies and dietary interventions may give us very useful information, especially to validate in the closest to human system what has been previously discovered in simpler organisms. To this purpose, three major studies have been performed in non-human primates to evaluate the efficacy and safety of calorie restriction and aging: one was performed at the University of Maryland [126], one was at NIA (National institute of Aging) [127], and the last one was at the Wisconsin National Primate Research Center [128]. All of them confirmed the benefits of CR on health. The University of Wisconsin study confirmed that CR reduces disease incidence and prolongs life span [129–131], while the NIA study doesn't observe a different survival between CR and control monkeys. Differences in feeding protocol, diet composition, and the timing of CR onset may explain the observed differences [132]. However, data from both studies suggest that many of the beneficial effects of CR reported in rodents also occur in primate models, thus suggesting a possible role also in human aging [131].

From comparative genetics studies between non-human primates and humans, it is deducible that some of the genetic regulatory processes that are important during development are less subjected to selective pressure and may became adverse later in life [133]. Blalock et al. analyzed gene expression in two major hippocampal subdivisions that are critical for memory/cognitive function in rhesus monkeys, identifying genes that changed expression with aging, and showed that increased gene expression of the glucocorticoid receptor happens with aging in rhesus macaques, linking the age-dependent metabolic syndrome to aging changes in the brain [134]. Epigenetic changes occur during aging in monkeys. In rhesus macaque brains, aging is associated with a global increase in H3K4me2 and H3K4me3 transcriptional activation; in addition, SETD7 and DPY30 (H3K4me2 methyltransferases) show elevated expression [135]. A role in aging has been ascribed to miRNA; in particular, Mohan et al. identified the miR-34a-SIRT1-acetylp65 axis as a potential mediator of "inflammaging" in the intestine [136].

8. Discussion

Humans are not an easy-to-study system both for ethical as well as for practical reasons, and have required the support of simpler organisms to model their physiology and pathology. Epidemiology, one of the most used approaches to investigate humans, is the branch of science that tries to count how often a certain pathology occurs in different population groups, hopefully identifying the risk or protective factors against certain pathologies. It is an essential methodology to study humans, but can fall short because it often requires a very large cohort of people, is normally very time and budget consuming, and can potentially be affected by many confounding factors. Clinical trials often use small groups of people to be cost effective and test specific hypotheses ranging from the safety/effectiveness of selected drugs to specific devices. Of course, obvious ethical as well as practical issues limit the potential of this approach in humans. Therefore, although both approaches are essential for the understanding of human physiology and pathology, they unlikely may suggest alone new branches of science or help to develop breakthrough ideas [137]. On the other hand, human genetics has been very successful, and revealed the genetic cause of more than 6000 monogenic diseases in humans, but it is hard to imagine the development of this science without the seminal discoveries in plants, fruitfly, fungi, etc., also in humans. The identification of genes and pathways relevant for human longevity has acquired deep advantages from model systems and discoveries, which can be grouped in four functional categories:

- Genes related to stress resistance: their role in longevity has first been demonstrated in many different model systems [18,24,25,32–34,51,54,58,82–85] and eventually confirmed in centenarians who show a low degree of oxidative stress as well as high antioxidant protection [138,139]. A high level of oxidative stress is also an important risk factor of other age-related diseases such as hypertension, atherosclerosis, and diabetes. SNP (single nucleotide polymorphisms) studies have identified Tp53, coding for tumor suppressor p53 [140–142], EXO1 [143], GPX1 (glutathione peroxidase1) [144], SOD2 (manganese superoxide dismutase) [145], heat shock proteins genes HSPA1A, HSPA1B, and HSPA1L [146–148], GSTZ1 (glutathione S-transferase zeta 1) [149], NOS1, NOS2 (nitric oxide synthase 1 and 2) [150], and UCPs (uncoupling proteins) [147,151,152] as susceptibility genes.
- Genes involved in telomeres length: they have been found to be associated with human longevity such as TERT and TERC (telomerase reverse transcriptase, telomerase RNA component) [153], SIRT1, and SIRT3 (sirtuins) [154,155]. The first discoveries were made in yeasts and tetrahymena by Elizabeth Blackburn, finding the role of TERT and TERC ([156] and references within). In *Caenorhabditis elegans* over-expressing a protein involved in telomere length regulation leads to the elongation of telomeres and extends the life span, making the organism more resistant to heat stress [157]. The over-expression of TERT also extends the life span of mice [158]. In yeast, sirtuins promote longevity [159]; in particular, it has been reported that Sir2 mediates life-span extension due by calorie restriction [160]. These findings have been replicated in other model organisms [161], but their role in longevity is not consistent for all species, and therefore is still under debate [162].
- Genes involved in metabolism and cellular division: APOE (apolipoprotein E) [163], TXNRD1 (thioredoxin reductase 1), XDH (xanthine dehydrogenase) [163], MAP3K7 (mitogen-activated protein kinase kinase kinase 7) [149], AKT kinase, and TOR [164]. The association of APOE with human longevity have been replicated in different populations: [165–167]. Apolipoprotein E (apoE) exhibits three isoforms: apoE2, apoE3 and apoE4. They are involved in inflammation, elevated lipid levels, and oxidative stress; furthermore, these are risk factors for cardiovascular disease and Alzheimer's disease, as reported by Huebbe et al. (2011) [168]. APOE2 has been defined as a longevity gene for its putative protective function; it is abundant in long-lived people, while APOE4, that differs from e3 allele at a single aa (112cys), and has been considered a frailty allele [169]. In fact, it increases the risk of Alzheimer's disease and cardiovascular diseases, maybe for a putative interaction with the *β amyloid* protein, and it is almost absent in centenarians.
- Genes belonging to the IGF/GH and insulin pathway: mutations in genes belonging to the insulin or insulin-like signaling pathway extend the life span of *Caenorhabditis elegans* [170,171], *Drosophila melanogaster* [59,109,172], and mice [69,173]. In humans, it has been observed that insulin sensitivity normally decreases during aging. On the other hand, centenarians are more sensitive to insulin than other people, and often show lower IGF-1 plasma levels [174]. SNP studies have found an association of particular alleles or haplotypes for INS (insulin) [175], INSR (insulin receptor) [176], IGF1 (insulin growth factor 1) [177], IGF1R (insulin growth factor 1 receptor); in fact, a specific haplotype of the IGF-I receptor and the kinase PI3KCB is frequently found in individuals living longer together with low plasma levels of IGF-1 [178], IGF2 (insulin growth factor 2) [179], IGF2R (insulin growth factor 2 receptor) [180], IRS1 (insulin receptor substrate 1) [177], GH1 (growth hormone 1) [177], GHSR (growth hormone secretagogue receptor type 1) [175], FOXO1A (forkhead box protein O1 A), and FOXO3A (forkhead box protein O3 A) transcription factor, which contains alleles that are associated with longevity in multiple Asian and European populations [181–185].

FOXO is a transcription factor that is conserved in all eukaryotic organisms and is negatively regulated by the insulin-signaling pathway. When insulin or insulin-like growth factor signaling is low, FOXO is activated, and life-span extension occurs [182].

Studies on all of the model organisms cited above have thus contributed to the discovery of the fundamental mechanisms of aging in humans. The essential conservation of these mechanisms throughout evolution has been strikingly confirmed in all the model organisms that have been tested so far. The series of similarities found in the mechanisms of regulation of aging in all the model organisms and in humans make us believe that these mechanisms have been preserved during the evolution from yeast to mammals [36,186]. For example, the protein sequence of DAF-2 (the *C. elegans* insulin/IGF receptor ortholog) shows 34% identity with the IGF-IR of mammals, nematode's AGE-1 is 27% identical to its ortholog PI3KCB kinase, and DAF-16 is 49% identical to FOXO1A, while IRS-1 has 30% identity with *Drosophila*'s insulin receptor substrate 1 (CHICO). In addition, these factors regulate similar processes in all organisms such as resistance to oxidative stress, metabolism, nutrient utilization, and of course life span [187]. A 2007 study compared genes whose transcription varies with the inhibition of the insulin/IGF-1 pathway in three different species—*C. elegans*, *Drosophila*, and mouse—and it was demonstrated that there are significant similarities concerning in particular two main categories of genes. The first one includes genes involved in protein synthesis that are hypoexpressed (this was also independently observed in yeast [188]), the second one includes genes involved in detoxification that are over-expressed, (e.g., the gene coding for glutathione S-transferase [189]). Consistent with the latter observation, the over-expression of transcription factors that regulate xenobiotics metabolism increases survival in *Caenorhabditis* and *Drosophila* [190]. The considerable affinities that were found confirm that the mechanism of aging, which is specifically mediated by the insulin-dependent pathway, has been preserved during the evolution in all eukaryotes from yeast to humans. In addition, it has been also confirmed that these pathways may be regulated mainly by nutrients [191]. Calorie restriction, protein restriction, fasting, and a fasting-mimicking diet are in fact becoming interesting alternatives to manipulating our genome structure [192]. Again, it has been so far confirmed that simple model systems, despite their clear morphological differences with humans, can be effectively used as a model to pave the way to future relevant discoveries in humans. Therefore, we believe that model systems will continue to be an essential tool for aging research and for the usage of high-throughput methodologies However, the observation that wild-caught animals may behave differently than lab animals stresses the need to confirm the results obtained with one organism in other species or in wild-caught animals in order to avoid the possibility that lab condition and repeated breeding may have favored genetic drift or epigenetic changes.

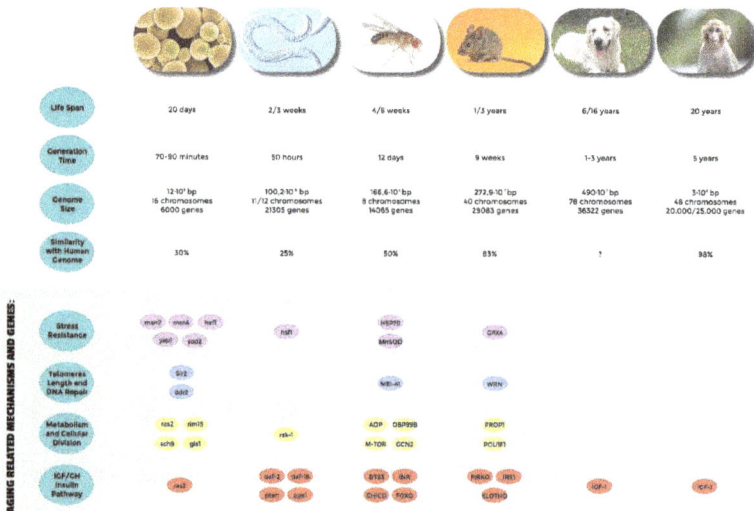

Figure 1. Comparative analysis of the most used model systems in aging research. Genome informations are from NCBI.

Author Contributions: G.T. wrote the paper; F.F. prepared the figure; S.V., N.G., A.R. revised and made suggestions to improve the paper; M.G.M. conceived, wrote andrevised the paper.

Funding: M.G.M. is funded by Create Cures foundation and The Laurus Project foundation.

Acknowledgments: We would like to thank Valentino Bellini for graphical assistance.

Conflicts of Interest: The authors declare no conflict of interest. The funders had no role in the design of the study; in the collection, analyses, or interpretation of data; in the writing of the manuscript, and in the decision to publish the results.

References

1. Williams, G.C. Pleiotropy, natural selection, and the evolution of senescence. *Evolution* **1957**, *11*, 398. [CrossRef]
2. Longo, V.D.; Finch, C.E. Evolutionary Medicine: From Dwarf Model Systems to Healthy Centenarians? *Science* **2003**, *299*, 1342. [CrossRef] [PubMed]
3. Kirkwood, T.B. Understanding the odd science of aging. *Cell* **2005**, *120*, 437. [CrossRef] [PubMed]
4. Longo, V.D.; Mitteldorf, J.; Skulachev, V.P. Programmed and altruistic ageing. *Nat. Rev. Genet.* **2005**, *6*, 866. [CrossRef] [PubMed]
5. Hartwell, L.H. Nobel Lecture. Yeast and cancer. *Biosci. Rep.* **2002**, *22*, 373–394. [CrossRef] [PubMed]
6. Coughlan, C.M.; Brodsky, J.L. Use of yeast as a model system to investigate protein confor-mational diseases. *Mol. Biotechnol.* **2005**, *30*, 171–180. [CrossRef]
7. Nakano, A. Yeast Golgi apparatus–Dynamics and sorting. *Cell Mol. Life Sci.* **2004**, *61*, 186–191. [CrossRef]
8. Bowers, K.; Stevens, T.H. Protein transport from the late Golgi to the vacuole in the yeast Sac-charomyces cerevisiae. *Biochim. Biophys. Acta* **2005**, *1744*, 438–454. [CrossRef]
9. Petranovic, D.; Nielsen, J. Can yeast systems biology contribute to the understanding of human disease? *Trends Biotechnol.* **2008**, *26*, 584–590. [CrossRef]
10. Karathia, H.; Vilaprinyo, E.; Sorribas, A.; Alves, R. Saccharomyces cerevisiae as a model or-ganism: A comparative study. *PLoS ONE* **2011**, *6*, e16015. [CrossRef]
11. Oliveira, A.V.; Vilaça, R.; Santos, C.N.; Costa, V.; Menezes, R. Exploring the power of yeast to model aging and age-related neurodegenerative disorders. *Biogerontology* **2017**, *18*, 3–34. [CrossRef] [PubMed]
12. Mirisola, M.G.; Braun, R.J.; Petranovic, D. Approaches to study yeast cell aging and death. *FEMS Yeast Res.* **2014**, *14*, 109–118. [CrossRef] [PubMed]
13. Longo, V.D.; Shadel, G.S.; Kaeberlein, M.; Kennedy, B. Replicative and chronological aging in Saccharomyces cerevisiae. *Cell Metab.* **2012**, *16*, 18–31. [CrossRef] [PubMed]
14. Mortimer, R.K.; Johnston, J.R. Life span of individual yeast cells. *Nature* **1959**, *183*, 1751–1752. [CrossRef] [PubMed]
15. Hu, J.; Wei, M.; Mirisola, M.G.; Longo, V.D. Assessing chronological aging in Saccharomyces cerevisiae. *Methods Mol. Biol.* **2013**, *965*, 463–472. [PubMed]
16. Mirisola, M.G.; Longo, V.D. Acetic acid and acidification accelerate chronological and replicative aging in yeast. *Cell Cycle* **2012**, *11*, 3532–3533. [CrossRef] [PubMed]
17. Fabrizio, P.; Gattazzo, C.; Battistella, L.; Wei, M.; Cheng, C.; McGrew, K.; Longo, V.D. Sir2 blocks extreme life-span extension. *Cell* **2005**, *123*, 655–667. [CrossRef]
18. Longo, V.D. Mutations in signal transduction proteins increase stress resistance and longevity in yeast, nematodes, fruit flies, and mammalian neuronal cells. *Neurobiol. Aging* **1999**, *20*, 479–486. [CrossRef]
19. Longo, V.D.; Gralla, E.B.; Valentine, J.S. Superoxide dismutase activity is essential for sta-tionary phase survival in Saccharomyces cerevisiae. Mitochondrial production of toxic oxygen spe-cies in vivo. *J. Biol. Chem.* **1996**, *271*, 12275–12280. [CrossRef]
20. Dhirendra, K.S.; Dwight, V.N.; McCormick, F. RAS Proteins and Their Regulators in Human Disease. *Cell* **2017**, *170*, 17–33.
21. Mirisola, M.G.; Seidita, G.; Verrotti, A.C.; Di Blasi, F.; Fasano, O. Mutagenic alteration of the distal switch II region of RAS blocks CDC25-dependent signaling functions. *J. Biol. Chem.* **1994**, *269*, 15740–15748. [PubMed]
22. Liu, Y.; Yang, F.; Li, S.; Dai, J.; Deng, H. Glutaredoxin Deletion Shortens Chronological Life Span in Saccharomyces cerevisiae via ROS-Mediated Ras/PKA Activation. *J. Proteome Res.* **2018**, *17*, 2318–2327. [CrossRef] [PubMed]

23. Pedruzzi, I.; Burckert, N.; Egger, P.; De Virgilio, C. Saccharomyces cerevisiae Ras/cAMP pathway controls post-diauxic shift element-dependent transcription through the zinc finger protein Gis1. *EMBO J.* **2000**, *19*, 2569–2579. [CrossRef] [PubMed]
24. Longo, V.D. The Chronological Life Span of Saccharomyces Cerevisiae. Studies of Superoxide Dismutase, Ras and Bcl-2. Ph.D. Thesis, University of California Los Angeles, Westwood, CA, USA, 1997.
25. Wei, M.; Fabrizio, P.; Hu, J.; Ge, H.; Cheng, C.; Li, L.; Longo, V.D. Life span extension by calorie restriction depends on Rim15 and transcription factors downstream of Ras/PKA, Tor, and Sch9. *PLoS Genet.* **2008**, *4*, e13. [CrossRef] [PubMed]
26. Martinez-Pastor, M.T.; Marchler, G.; Schuller, C.; Marchler-Bauer, A.; Ruis, H.; Estruch, F. The Saccharomyces cerevisiae zinc finger proteins Msn2p and Msn4p are required for transcriptional induction through the stress response element (STRE). *EMBO J.* **1996**, *15*, 2227–2235. [CrossRef]
27. Fabrizio, P.; Pozza, F.; Pletcher, S.D.; Gendron, C.M.; Longo, V.D. Regulation of longevity and stress resistance by Sch9 in yeast. *Science* **2001**, *292*, 288–290. [CrossRef]
28. Hu, J.; Wei, M.; Mirzaei, H.; Madia, F.; Mirisola, M.; Amparo, C.; Chagoury, S.; Kennedy, B.; Longo, V.D. Tor-Sch9 deficiency activates catabolism of the ketone body-like acetic acid to promote trehalose accumulation and longevity. *Aging Cell* **2014**, *13*, 457–467. [CrossRef]
29. Toda, T.; Cameron, S.; Sass, P.; Wigler, M. SCH9, a gene of Saccharomyces cerevisiae that encodes a protein distinct from, but functionally and structurally related to, cAMP-dependent protein kinase catalytic subunits. *Genes Dev.* **1988**, *2*, 517–527. [CrossRef]
30. Thevelein, J.M.; De Winde, J.H. Novel sensing mechanisms and targets for the cAMP-protein kinase A pathway in the yeast Saccharomyces cerevisiae. *Mol. Microbiol.* **1999**, *33*, 904–918. [CrossRef]
31. Mirisola, M.G.; Taormina, G.; Fabrizio, P.; Wei, M.; Hu, J.; Longo, V.D. Serine- and threo-nine/valine-dependent activation of PDK and Tor orthologs converge on Sch9 to promote aging. *PLoS Genet.* **2014**, *10*, e1004113. [CrossRef]
32. Flattery-O'Brien, J.A.; Grant, C.M.; Dawes, I.W. Stationary-phase regulation of the Saccha-romyces cerevisiaeSOD2 gene is dependent on additive effects of HAP2/3/4/5- and STRE-binding elements. *Mol. Microbiol.* **1997**, *23*, 303–312. [CrossRef] [PubMed]
33. Harris, N.; MacLean, M.; Hatzianthis, K.; Panaretou, B.; Piper, P.W. Increasing Saccharomyces cerevisiae stress resistance, through the overactivation of the heat shock response resulting from defects in the Hsp90 chaperone, does not extend replicative life span but can be associated with slower chronological ageing of nondividing cells. *Mol. Genet. Genom.* **2001**, *265*, 258–263.
34. Herker, E.; Jungwirth, H.; Lehmann, K.A.; Maldener, C.; Frohlich, K.U.; Wissing, S.; Buttner, S.; Fehr, M.; Sigrist, S.; Madeo, F. Chronological aging leads to apoptosis in yeast. *J. Cell Biol.* **2004**, *164*, 501–507. [CrossRef] [PubMed]
35. Longo, V.D. The Ras and Sch9 pathways regulate stress resistance and longevity. *Exp. Gerontol.* **2003**, *38*, 807–811. [CrossRef]
36. Fontana, L.; Partridge, L.; Longo, V.D. Extending healthy life span–From yeast to humans. *Science* **2010**, *328*, 321–326. [CrossRef] [PubMed]
37. Parrella, E.; Longo, V.D. Insulin/IGF-I and related signaling pathways regulate aging in nondividing cells: From yeast to the mammalian brain. *Sci. World J.* **2010**, *10*, 161–177. [CrossRef] [PubMed]
38. Xu, X.; Kim, S.K. The early bird catches the worm: New technologies for the Caenorhabditis elegans toolkit. *Nat. Rev. Genet.* **2011**, *12*, 793–801. [CrossRef] [PubMed]
39. Boulin, T.; Hobert, O. From genes to function: The *C. elegans* genetic toolbox. *Wiley Interdiscip. Rev. Dev. Biol.* **2012**, *1*, 114–137. [CrossRef] [PubMed]
40. Lai, C.H.; Chou, C.Y.; Ch'ang, L.Y.; Liu, C.S.; Lin, W. Identification of novel human genes evolutionarily conserved in *Caenorhabditis elegans* by comparative proteomics. *Genome Res.* **2000**, *10*, 703–713. [CrossRef]
41. Calixto, A.; Ma, C.; Chalfie, M. Conditional gene expression and RNAi using MEC-8-dependent splicing in *C. elegans*. *Nat. Methods* **2010**, *7*, 407–411. [CrossRef]
42. Tarkhov, A.E.; Alla, R.; Ayyadevara, S.; Pyatnitskiy, M.; Menshikov, L.I.; Reis, R.J.S.; Fedichev, P.O. A universal transcriptomic signature of age reveals the temporal scaling of *Caenorhabditis elegans* aging trajectories. *Sci. Rep.* **2019**, *9*, 7368. [CrossRef] [PubMed]
43. Son, H.G.; Altintas, O.; Kim, E.J.E.; Kwon, S.; Lee, S.V. Age-dependent changes and biomarkers of aging in *Caenorhabditis elegans*. *Aging Cell* **2019**, *18*, e12853. [CrossRef] [PubMed]

44. Kenyon, C.; Chang, J.; Gensch, E.; Rudner, A.; Tabtiang, R.A.C. *C. elegans* mutant that lives twice as long as wild type. *Nature* **1993**, *366*, 461–464. [CrossRef] [PubMed]

45. Riddle, D.L.; Albert, P.S. Genetic and Environmental Regulation of Dauer Larva Development. In *C. elegans II*, 2nd ed.; Riddle, D.L., Blumenthal, T., Meyer, B.J., Priess, J.R., Eds.; Cold Spring Harbor Laboratory Press: New York, NY, USA, 1997.

46. Kimura, K.D.; Tissenbaum, H.A.; Liu, Y.; Ruvkun, G. *daf-2*, an insulin receptor-like gene that regulates longevity and diapause in *Caenorhabditis elegans*. *Science* **1997**, *277*, 942–946. [CrossRef] [PubMed]

47. Lin, K.; Dorman, J.B.; Rodan, A.; Kenyon, C. *daf-16*: An HNF-3/forkhead family member that can function to double the life-span of *Caenorhabditis elegans*. *Science* **1997**, *278*, 1319–1322. [CrossRef] [PubMed]

48. Henderson, S.T.; Johnson, T.E. *daf-16* integrates developmental and environmental inputs to mediate aging in the nematode *Caenorhabditis elegans*. *Curr. Biol.* **2001**, *11*, 1975–1980. [CrossRef]

49. Lee, R.Y.; Hench, J.; Ruvkun, G. Regulation of *C. elegans* DAF-16 and its human ortholog FKHRL1 by the *daf-2* insulin-like signaling pathway. *Curr. Biol.* **2001**, *11*, 1950–1957. [CrossRef]

50. Hesp, K.; Smant, G.; Kammenga, J.E. Caenorhabditis elegans DAF-16/FOXO transcription factor and its mammalian homologs associate with age-related disease. *Exp. Gerontol.* **2015**, *72*, 1–7. [CrossRef]

51. Morley, J.F.; Morimoto, R.I. Regulation of longevity in *Caenorhabditis elegans* by heat shock factor and molecular chaperones. *Mol. Biol. Cell* **2004**, *15*, 657–664. [CrossRef]

52. Larsen, P.L.; Albert, P.S.; Riddle, D.L. Genes that regulate both development and longevity in *Caenorhabditis elegans*. *Genetics* **1995**, *139*, 1567–1583.

53. Dorman, J.B.; Albinder, B.; Shroyer, T.; Kenyon, C. The *age-1* and *daf-2* genes function in a common pathway to control the lifespan of *Caenorhabditis elegans*. *Genetics* **1995**, *141*, 1399–1406. [PubMed]

54. Honda, Y.; Honda, S. Oxidative stress and life span determination in the nematode *Caenorhabditis elegans*. *Ann. N. Y. Acad. Sci.* **2002**, *959*, 466–474. [CrossRef] [PubMed]

55. Branicky, R.; Benard, C.; Hekimi, S. *clk-1*, mitochondria, and physiological rates. *Bioessays* **2000**, *22*, 48–56. [CrossRef]

56. Piper, M.D.W.; Partridge, L. Drosophila as a model for ageing. *Biochim. Biophys. Acta Mol. Basis Dis.* **2018**, *1864*, 2707–2717. [CrossRef] [PubMed]

57. Brandt, A.; Vilcinskas, A. The Fruit Fly *Drosophila melanogaster* as a Model for Aging Research. *Adv. Biochem. Eng. Biotechnol.* **2013**, *135*, 63–77. [PubMed]

58. Orr, W.C.; Sohal, R.S. Extension of life-span by overexpression of superoxide dismutase and catalase in *Drosophila melanogaster*. *Science* **1994**, *263*, 1128–1130. [CrossRef]

59. Clancy, D.J.; Gems, D.; Harshman, L.G.; Oldham, S.; Stocker, H.; Hafen, E.; Leevers, S.J.; Partridge, L. Extension of life-span by loss of CHICO, a Drosophila insulin receptor substrate protein. *Science* **2001**, *292*, 104–106. [CrossRef]

60. Sun, J.; Tower, J. FLP recombinase-mediated induction of Cu/Zn-superoxide dismutase transgene expression can extend the life span of adult *Drosophila melanogaster* flies. *Mol. Cell Biol.* **1999**, *19*, 216–228. [CrossRef]

61. Piper, M.D.; Selman, C.; McElwee, J.J.; Partridge, L. Separating cause from effect: How does insulin/IGF signalling control lifespan in worms, flies and mice? *J. Intern. Med.* **2008**, *263*, 179–191. [CrossRef]

62. Alic, N.; Giannakou, M.E.; Papatheodorou, I.; Hoddinott, M.P.; Andrews, T.D.; Bolukbasi, E.; Partridge, L. Interplay of dFOXO and two ETS-family transcription factors determines lifespan in *Drosophila melanogaster*. *PLoS Genet.* **2014**, *10*, e1004619. [CrossRef]

63. Kucerova, L.; Kubrak, O.I.; Bengtsson, J.M.; Strnad, H.; Nylin, S.; Theopold, U.; Nassel, D.R. Slowed aging during reproductive dormancy is reflected in genome-wide transcriptome changes in *Drosophila melanogaster*. *BMC Genom.* **2016**, *17*, 50. [CrossRef]

64. Bjedov, I.; Toivonen, J.M.; Kerr, F.; Slack, C.; Jacobson, J.; Foley, A.; Partridge, L. Mechanisms of life span extension by rapamycin in the fruit fly *Drosophila melanogaster*. *Cell Metab.* **2010**, *11*, 35–46. [CrossRef]

65. Kapahi, P.; Zid, B.M.; Harper, T.; Koslover, D.; Sapin, V.; Benzer, S. Regulation of lifespan in Drosophila by modulation of genes in the TOR signaling pathway. *Curr. Biol.* **2004**, *14*, 885–890. [CrossRef]

66. Katewa, S.D.; Kapahi, P. Role of TOR signaling in aging and related biological processes in *Drosophila melanogaster*. *Exp. Gerontol.* **2011**, *46*, 382–390. [CrossRef]

67. Ulgherait, M.; Rana, A.; Rera, M.; Graniel, J.; Walker, D.W. AMPK modulates tissue and or-ganismal aging in a non-cell-autonomous manner. *Cell Rep.* **2014**, *8*, 1767–1780. [CrossRef]

68. Chapman, T.; Partridge, L. Female fitness in *Drosophila melanogaster*: An interaction between the effect of nutrition and of encounter rate with males. *Proc. Biol. Sci.* **1996**, *263*, 755–759.

69. Kang, M.J.; Vasudevan, D.; Kang, K.; Kim, K.; Park, J.E.; Zhang, N.; Zeng, X.; Neubert, T.A.; Marr, M.T., 2nd; Ryoo, H.D. 4E-BP is a target of the GCN2-ATF4 pathway during Drosophila development and aging. *J. Cell Biol.* **2017**, *216*, 115–129. [CrossRef]

70. Brown-Borg, H.M.; Borg, K.E.; Meliska, C.J.; Bartke, A. Dwarf mice and the ageing process. *Nature* **1996**, *384*, 33. [CrossRef]

71. Dolle, M.E.; Snyder, W.K.; Vijg, J. Genotyping the *Prop-1* mutation in Ames dwarf mice. *Mech. Ageing Dev.* **2001**, *122*, 1915–1918. [CrossRef]

72. Andersen, B.; Pearse, R.V., 2nd; Jenne, K.; Sornson, M.; Lin, S.C.; Bartke, A.; Rosenfeld, M.G. The Ames dwarf gene is required for *Pit-1* gene activation. *Dev. Biol.* **1995**, *172*, 495–503. [CrossRef]

73. Snell, G.D. Dwarf, a new mendelian recessive character of the house mouse. *Proc. Natl. Acad. Sci. USA* **1929**, *15*, 733–734. [CrossRef]

74. Flurkey, K.; Papaconstantinou, J.; Miller, R.A.; Harrison, D.E. Lifespan extension and delayed immune and collagen aging in mutant mice with defects in growth hormone production. *Proc. Natl. Acad. Sci. USA* **2001**, *98*, 6736–6741. [CrossRef]

75. Flurkey, K.; Papaconstantinou, J.; Harrison, D.E. The Snell dwarf mutation Pit1(dw) can in-crease life span in mice. *Mech. Ageing Dev.* **2002**, *123*, 121–130. [CrossRef]

76. Turyn, D.; Dominici, F.P.; Sotelo, A.I.; Bartke, A. Specific interactions of growth hormone (GH) with GH-receptors and GH-binding proteins in vivo in genetically GH-deficient Ames dwarf mice. *Growth Horm. IGF Res.* **1998**, *8*, 389–396. [CrossRef]

77. Bartke, A.; Brown-Borg, H.M.; Bode, A.M.; Carlson, J.; Hunter, W.S.; Bronson, R.T. Does growth hormone prevent or accelerate aging? *Exp. Gerontol.* **1998**, *33*, 675–687. [CrossRef]

78. Hunter, W.S.; Croson, W.B.; Bartke, A.; Gentry, M.V.; Meliska, C.J. Low body temperature in long-lived Ames dwarf mice at rest and during stress. *Physiol. Behav.* **1999**, *67*, 433–437. [CrossRef]

79. Svare, B.; Bartke, A.; Doherty, P.; Mason, I.; Michael, S.D.; Smith, M.S. Hyperprolactinemia suppresses copulatory behavior in male rats and mice. *Biol. Reprod.* **1979**, *21*, 529–535. [CrossRef]

80. Garcia, A.M.; Busuttil, R.A.; Calder, R.B.; Dolle, M.E.; Diaz, V.; McMahan, C.A.; Bartke, A.; Nelson, J.; Reddick, R.; Vijg, J. Effect of Ames dwarfism and caloric restriction on spontaneous DNA mutation frequency in different mouse tissues. *Mech. Ageing Dev.* **2008**, *129*, 528–533. [CrossRef]

81. Alderman, J.M.; Flurkey, K.; Brooks, N.L.; Naik, S.B.; Gutierrez, J.M.; Srinivas, U.; Ziara, K.B.; Jing, L.; Boysen, G.; Bronson, R.; et al. Neuroendocrine inhibition of glucose production and resistance to cancer in dwarf mice. *Exp. Gerontol.* **2009**, *44*, 26–33. [CrossRef]

82. Brown-Borg, H.; Johnson, W.T.; Rakoczy, S.; Romanick, M. Mitochondrial oxidant gen-eration and oxidative damage in Ames dwarf and GH transgenic mice. *J. Am. Aging Assoc.* **2001**, *24*, 85–96.

83. Brown-Borg, H.M.; Bode, A.M.; Bartke, A. Antioxidative mechanisms and plasma growth hormone levels: Potential relationship in the aging process. *Endocrine* **1999**, *11*, 41–48. [CrossRef]

84. Brown-Borg, H.M.; Rakoczy, S.G. Catalase expression in delayed and premature aging mouse models. *Exp. Gerontol.* **2000**, *35*, 199–212. [CrossRef]

85. Bokov, A.F.; Lindsey, M.L.; Khodr, C.; Sabia, M.R.; Richardson, A. Long-lived ames dwarf mice are resistant to chemical stressors. *J. Gerontol. A Biol. Sci. Med. Sci.* **2009**, *64*, 819–827. [CrossRef]

86. Patrick, A.; Seluanov, M.; Hwang, C.; Tam, J.; Khan, T.; Morgenstern, A.; Wiener, L.; Vazquez, J.M.; Zafar, H.; Wen, R.; et al. Sensitivity of primary fibroblasts in culture to atmospheric oxygen does not correlate with species lifespan. *Aging* **2016**, *8*, 841–847. [CrossRef]

87. Bluher, M.; Kahn, B.B.; Kahn, C.R. Extended longevity in mice lacking the insulin receptor in adipose tissue. *Science* **2003**, *299*, 572–574. [CrossRef]

88. Bartke, A. Minireview: Role of the growth hormone/insulin-like growth factor system in mammalian aging. *Endocrinology* **2005**, *146*, 3718–3723. [CrossRef]

89. Selman, C.; Tullet, J.M.; Wieser, D.; Irvine, E.; Lingard, S.J.; Choudhury, A.I.; Claret, M.; Al-Qassab, H.; Carmignac, D.; Ramadani, F.; et al. Ribosomal protein S6 kinase 1 signaling regulates mammalian life span. *Science* **2009**, *326*, 140–144. [CrossRef]

90. Kuro-o, M.; Matsumura, Y.; Aizawa, H.; Kawaguchi, H.; Suga, T.; Utsugi, T.; Ohyama, Y.; Kurabayashi, M.; Kaname, T.; Kume, E.; et al. Mutation of the mouse klotho gene leads to a syndrome resembling ageing. *Nature* **1997**, *390*, 45–51. [CrossRef]
91. Kurosu, H.; Yamamoto, M.; Clark, J.D.; Pastor, J.V.; Nandi, A.; Gurnani, P.; McGuinness, O.P.; Chikuda, H.; Yamaguchi, M.; Kawaguchi, H.; et al. Suppression of aging in mice by the hormone Klotho. *Science* **2005**, *309*, 1829–1833. [CrossRef]
92. Bektas, A.; Schurman, S.H.; Sharov, A.A.; Carter, M.G.; Dietz, H.C.; Francomano, C.A. Klotho gene variation and expression in 20 inbred mouse strains. *Mamm. Genome* **2004**, *15*, 759–767. [CrossRef]
93. Chang, S. A mouse model of Werner Syndrome: What can it tell us about aging and cancer? *Int. J. Biochem. Cell Biol.* **2005**, *37*, 991–999. [CrossRef]
94. Austad, S.N. Comparative biology of aging. *J. Gerontol. A Biol. Sci. Med. Sci.* **2009**, *64*, 199–201. [CrossRef]
95. Greer, K.A.; Canterberry, S.C.; Murphy, K.E. Statistical analysis regarding the effects of height and weight on life span of the domestic dog. *Res. Vet. Sci.* **2007**, *82*, 208–214. [CrossRef]
96. Bonnett, B.N.; Egenvall, A. Age patterns of disease and death in insured Swedish dogs, cats and horses. *J. Comp. Pathol.* **2010**, *142*, S33–S38. [CrossRef]
97. Fleming, J.M.; Creevy, K.E.; Promislow, D.E. Mortality in north american dogs from 1984 to 2004: An investigation into age-, size-, and breed-related causes of death. *J. Vet. Intern. Med.* **2011**, *25*, 187–198. [CrossRef]
98. Creevy, K.E.; Austad, S.N.; Hoffman, J.M.; O'Neill, D.G.; Promislow, D.E. The Companion Dog as a Model for the Longevity Dividend. *Cold Spring Harb. Perspect. Med.* **2016**, *6*, a026633. [CrossRef]
99. Freeman, L.M. Cachexia and sarcopenia: Emerging syndromes of importance in dogs and cats. *J. Vet. Intern. Med.* **2012**, *26*, 3–17. [CrossRef]
100. Urfer, S.R.; Greer, K.; Wolf, N.S. Age-related cataract in dogs: A biomarker for life span and its relation to body size. *Age (Dordr)* **2011**, *33*, 451–460. [CrossRef]
101. Vite, C.H.; Head, E. Aging in the canine and feline brain. *Vet. Clin. N. Am. Small Anim. Pract.* **2014**, *44*, 1113–1129. [CrossRef]
102. Boyko, A.R.; Quignon, P.; Li, L.; Schoenebeck, J.J.; Degenhardt, J.D.; Lohmueller, K.E.; Zhao, K.; Brisbin, A.; Parker, H.G.; vonHoldt, B.M.; et al. A simple genetic architecture underlies morphological variation in dogs. *PLoS Biol.* **2010**, *8*, e1000451. [CrossRef]
103. Vonholdt, B.M.; Pollinger, J.P.; Lohmueller, K.E.; Han, E.; Parker, H.G.; Quignon, P.; De-genhardt, J.D.; Boyko, A.R.; Earl, D.A.; Auton, A.; et al. Genome-wide SNP and haplotype analyses reveal a rich history underlying dog domestication. *Nature* **2010**, *464*, 898–902. [CrossRef] [PubMed]
104. Ostrander, E.A.; Wayne, R.K. The canine genome. *Genome Res.* **2005**, *15*, 1706–1716. [CrossRef] [PubMed]
105. Gilmore, K.M.; Greer, K.A. Why is the dog an ideal model for aging research? *Exp. Gerontol.* **2015**, *71*, 14–20. [CrossRef] [PubMed]
106. Al-Regaiey, K.A.; Masternak, M.M.; Bonkowski, M.; Sun, L.; Bartke, A. Long-lived growth hormone receptor knockout mice: Interaction of reduced insulin-like growth factor i/insulin signaling and caloric restriction. *Endocrinology* **2005**, *146*, 851–860. [CrossRef]
107. Eigenmann, J.E.; Amador, A.; Patterson, D.F. Insulin-like growth factor I levels in proportionate dogs, chondrodystrophic dogs and in giant dogs. *Acta Endocrinol. (Copenh)* **1988**, *118*, 105–108. [CrossRef]
108. Greer, K.A.; Hughes, L.M.; Masternak, M.M. Connecting serum IGF-1, body size, and age in the domestic dog. *Age (Dordr)* **2011**, *33*, 475–483. [CrossRef]
109. Tatar, M.; Bartke, A.; Antebi, A. The endocrine regulation of aging by insulin-like signals. *Science* **2003**, *299*, 1346–1351. [CrossRef]
110. Waters, D.J.; Kengeri, S.S.; Clever, B.; Booth, J.A.; Maras, A.H.; Schlittler, D.L.; Hayek, M.G. Exploring mechanisms of sex differences in longevity: Lifetime ovary exposure and exceptional lon-gevity in dogs. *Aging Cell* **2009**, *8*, 752–755. [CrossRef]
111. Adams, V.J.; Ceccarelli, K.; Watson, P.; Carmichael, S.; Penell, J.; Morgan, D.M. Evidence of longer life; a cohort of 39 labrador retrievers. *Vet. Rec.* **2018**, *182*, 408. [CrossRef]
112. Kealy, R.D.; Lawler, D.F.; Ballam, J.M.; Mantz, S.L.; Biery, D.N.; Greeley, E.H.; Lust, G.; Segre, M.; Smith, G.K.; Stowe, H.D. Effects of diet restriction on life span and age-related changes in dogs. *J. Am. Vet. Med. Assoc.* **2002**, *220*, 1315–1320. [CrossRef]

113. Lawler, D.F.; Larson, B.T.; Ballam, J.M.; Smith, G.K.; Biery, D.N.; Evans, R.H.; Greeley, E.H.; Segre, M.; Stowe, H.D.; Kealy, R.D. Diet restriction and ageing in the dog: Major observations over two decades. *Br. J. Nutr.* **2008**, *99*, 793–805. [CrossRef]

114. Richards, S.E.; Wang, Y.; Claus, S.P.; Lawler, D.; Kochhar, S.; Holmes, E.; Nicholson, J.K. Metabolic phenotype modulation by caloric restriction in a lifelong dog study. *J. Proteome Res.* **2013**, *12*, 3117–3127. [CrossRef]

115. Greeley, E.H.; Spitznagel, E.; Lawler, D.F.; Kealy, R.D.; Segre, M. Modulation of canine im-munosenescence by life-long caloric restriction. *Vet. Immunol. Immunopathol.* **2006**, *111*, 287–299. [CrossRef]

116. Jimenez, A.G.; Winward, J.; Beattie, U.; Cipolli, W. Cellular metabolism and oxidative stress as a possible determinant for longevity in small breed and large breed dogs. *PLoS ONE* **2018**, *13*, e0195832. [CrossRef]

117. Alexander, J.E.; Colyer, A.; Haydock, R.M.; Hayek, M.G.; Park, J. Understanding How Dogs Age: Longitudinal Analysis of Markers of Inflammation, Immune Function, and Oxidative Stress. *J. Gerontol. A Biol. Sci. Med. Sci.* **2018**, *73*, 720–728. [CrossRef]

118. Gibbs, R.A.; Rogers, J.; Katze, M.G.; Bumgarner, R.; Weinstock, G.M.; Mardis, E.R.; Remington, K.A.; Strausberg, R.L.; Venter, J.C.; Wilson, R.K.; et al. Evolutionary and biomedical insights from the rhesus macaque genome. *Science* **2007**, *316*, 222–234.

119. Zimin, A.V.; Cornish, A.S.; Maudhoo, M.D.; Gibbs, R.M.; Zhang, X.; Pandey, S.; Meehan, D.T.; Wipfler, K.; Bosinger, S.E.; Johnson, Z.P.; et al. A new rhesus macaque assembly and annotation for next-generation sequencing analyses. *Biol. Direct* **2014**, *9*, 20. [CrossRef]

120. Yates, A.; Akanni, W.; Amode, M.R.; Barrell, D.; Billis, K.; Carvalho-Silva, D.; Cummins, C.; Clapham, P.; Fitzgerald, S.; Gil, L.; et al. Ensembl 2016. *Nucleic Acids Res.* **2016**, *44*, D710–D716. [CrossRef]

121. Chimpanzee Sequencing and Analysis Consortium. Initial sequence of the chimpanzee genome and comparison with the human genome. *Nature* **2005**, *437*, 69–87. [CrossRef]

122. Colman, R.J. Non-human primates as a model for aging. *Biochim. Biophys. Acta Mol. Basis Dis.* **2018**, *1864*, 2733–2741. [CrossRef]

123. Uno, H. Age-related pathology and biosenescent markers in captive rhesus macaques. *Age (Omaha)* **1997**, *20*, 1–13. [CrossRef]

124. Colman, R.J.; McKiernan, S.H.; Aiken, J.M.; Weindruch, R. Muscle mass loss in Rhesus monkeys: Age of onset. *Exp. Gerontol.* **2005**, *40*, 573–581. [CrossRef]

125. Colman, R.J.; Kemnitz, J.W.; Lane, M.A.; Abbott, D.H.; Binkley, N. Skeletal effects of aging and menopausal status in female rhesus macaques. *J. Clin. Endocrinol. Metab.* **1999**, *84*, 4144–4148. [CrossRef]

126. Bodkin, N.L.; Alexander, T.M.; Ortmeyer, H.K.; Johnson, E.; Hansen, B.C. Mortality and morbidity in laboratory-maintained Rhesus monkeys and effects of long-term dietary restriction. *J. Gerontol. A Biol. Sci. Med. Sci.* **2003**, *58*, 212–219. [CrossRef]

127. Mattison, J.A.; Roth, G.S.; Beasley, T.M.; Tilmont, E.M.; Handy, A.M.; Herbert, R.L.; Longo, D.L.; Allison, D.B.; Young, J.E.; Bryant, M.; et al. Impact of caloric restriction on health and survival in rhesus monkeys from the NIA study. *Nature* **2012**, *489*, 318–321. [CrossRef]

128. Ramsey, J.J.; Colman, R.J.; Binkley, N.C.; Christensen, J.D.; Gresl, T.A.; Kemnitz, J.W.; Weindruch, R. Dietary restriction and aging in rhesus monkeys: The University of Wisconsin study. *Exp. Gerontol.* **2000**, *35*, 1131–1149. [CrossRef]

129. Colman, R.J.; Anderson, R.M.; Johnson, S.C.; Kastman, E.K.; Kosmatka, K.J.; Beasley, T.M.; Allison, D.B.; Cruzen, C.; Simmons, H.A.; Kemnitz, J.W.; et al. Caloric restriction delays disease onset and mortality in rhesus monkeys. *Science* **2009**, *325*, 201–204. [CrossRef]

130. Colman, R.J.; Beasley, T.M.; Kemnitz, J.W.; Johnson, S.C.; Weindruch, R.; Anderson, R.M. Caloric restriction reduces age-related and all-cause mortality in rhesus monkeys. *Nat. Commun.* **2014**, *5*, 3557. [CrossRef]

131. Mattison, J.A.; Colman, R.J.; Beasley, T.M.; Allison, D.B.; Kemnitz, J.W.; Roth, G.S.; Ingram, D.K.; Weindruch, R.; De Cabo, R.; Anderson, R.M. Caloric restriction improves health and survival of rhesus monkeys. *Nat. Commun.* **2017**, *8*, 14063. [CrossRef]

132. Vaughan, K.L.; Kaiser, T.; Peaden, R.; Anson, R.M.; De Cabo, R.; Mattison, J.A. Caloric Re-striction Study Design Limitations in Rodent and Nonhuman Primate Studies. *J. Gerontol. A Biol. Sci. Med. Sci.* **2017**, *73*, 48–53. [CrossRef]

133. De Magalhaes, J.P.; Church, G.M. Analyses of human-chimpanzee orthologous gene pairs to explore evolutionary hypotheses of aging. *Mech. Ageing Dev.* **2007**, *128*, 355–364. [CrossRef] [PubMed]

134. Blalock, E.M.; Grondin, R.; Chen, K.C.; Thibault, O.; Thibault, V.; Pandya, J.D.; Dowling, A.; Zhang, Z.; Sullivan, P.; Porter, N.M.; et al. Aging-related gene expression in hippocampus proper compared with dentate gyrus is selectively associated with metabolic syndrome variables in rhesus monkeys. *J. NeuroSci.* **2010**, *30*, 6058–6071. [CrossRef] [PubMed]

135. Han, Y.; Han, D.; Yan, Z.; Boyd-Kirkup, J.D.; Green, C.D.; Khaitovich, P.; Han, J.D. Stress-associated H3K4 methylation accumulates during postnatal development and aging of rhesus ma-caque brain. *Aging Cell* **2012**, *11*, 1055–1064. [CrossRef] [PubMed]

136. Mohan, M.; Kumar, V.; Lackner, A.A.; Alvarez, X. Dysregulated miR-34a-SIRT1-acetyl p65 axis is a potential mediator of immune activation in the colon during chronic simian immunodeficiency virus infection of rhesus macaques. *J. Immunol.* **2015**, *194*, 291–306. [CrossRef] [PubMed]

137. Taormina, G.; Mirisola, M.G. Longevity: Epigenetic and biomolecular aspects. *BioMol. Concepts* **2015**, *6*, 105–117. [CrossRef] [PubMed]

138. Paolisso, G.; Tagliamonte, M.R.; Rizzo, M.R.; Manzella, D.; Gambardella, A.; Varricchio, M. Oxidative stress and advancing age: Results in healthy centenarians. *J. Am. Geriatr. Soc.* **1998**, *46*, 833–838. [CrossRef]

139. Mecocci, P.; Polidori, M.C.; Troiano, L.; Cherubini, A.; Cecchetti, R.; Pini, G.; Straatman, M.; Monti, D.; Stahl, W.; Sies, H.; et al. Plasma antioxidants and longevity: A study on healthy centenarians. *Free Radic. Biol. Med.* **2000**, *28*, 1243–1248. [CrossRef]

140. Di Pietro, F.; Dato, S.; Carpi, F.M.; Corneveaux, J.J.; Serfaustini, S.; Maoloni, S.; Mignini, F.; Huentelman, M.J.; Passarino, G.; Napolioni, V. TP53*P72 allele influences negatively female life expectancy in a population of central Italy: Cross-sectional study and genetic-demographic approach analysis. *J. Gerontol. A Biol. Sci. Med. Sci.* **2013**, *68*, 539–545. [CrossRef]

141. Van Heemst, D.; Mooijaart, S.P.; Beekman, M.; Schreuder, J.; De Craen, A.J.; Brandt, B.W.; Slagboom, P.E.; Westendorp, R.G. Variation in the human TP53 gene affects old age survival and cancer mortality. *Exp. Gerontol.* **2005**, *40*, 11–15. [CrossRef]

142. Altilia, S.; Santoro, A.; Malagoli, D.; Lanzarini, C.; Álvarez, J.A.B.; Galazzo, G.; Porter, D.C.; Crocco, P.; Rose, G.; Passarino, G.; et al. TP53 codon 72 polymorphism affects accumulation of mtDNA damage in human cells. *Aging* **2012**, *4*, 28–39. [CrossRef]

143. Nebel, A.; Flachsbart, F.; Till, A.; Caliebe, A.; Blanché, H.; Arlt, A.; Häsler, R.; Jacobs, G.; Kleindorp, R.; Franke, A.; et al. A functional EXO1 promoter variant is associated with prolonged life expectancy in centenarians. *Mech. Ageing Dev.* **2009**, *130*, 691–699. [CrossRef]

144. Soerensen, M.; Christensen, K.; Stevnsner, T.; Christiansen, L. The Mn-superoxide dismutase single nucleotide polymorphism rs4880 and the glutathione peroxidase 1 single nucleotide poly-morphism rs1050450 are associated with aging and longevity in the oldest old. *Mech. Ageing Dev.* **2009**, *130*, 308–314. [CrossRef]

145. Lunetta, K.L.; D'Agostino, R.B., Sr.; Karasik, D.; Benjamin, E.J.; Guo, C.Y.; Govindaraju, R.; Kiel, D.P.; Kelly-Hayes, M.; Massaro, J.M.; Pencina, M.J.; et al. Genetic correlates of longevity and selected age-related phenotypes: A genome-wide association study in the Framingham Study. *BMC Med. Genet.* **2007**, *8*, S13. [CrossRef]

146. Singh, R.; Kolvraa, S.; Bross, P.; Christensen, K.; Gregersen, N.; Tan, Q.; Jensen, U.B.; Eiberg, H.; Rattan, S.I. Heat-shock protein 70 genes and human longevity: A view from Denmark. *Ann. N. Y. Acad. Sci.* **2006**, *1067*, 301–308. [CrossRef]

147. Ross, O.A.; Curran, M.D.; Crum, K.A.; Rea, I.M.; Barnett, Y.A.; Middleton, D. Increased frequency of the 2437T allele of the heat shock protein 70-Hom gene in an aged Irish population. *Exp. Gerontol.* **2003**, *38*, 561–565. [CrossRef]

148. Altomare, K.; Greco, V.; Bellizzi, D.; Berardelli, M.; Dato, S.; DeRango, F.; Garasto, S.; Rose, G.; Feraco, E.; Mari, V.; et al. The allele (A)(-110) in the promoter region of the HSP70-1 gene is unfavorable to longevity in women. *Biogerontology* **2003**, *4*, 215–220. [CrossRef]

149. Di Cianni, F.; Campa, D.; Tallaro, F.; Rizzato, C.; De Rango, F.; Barale, R.; Passarino, G.; Canzian, F.; Gemignani, F.; Montesanto, A.; et al. MAP3K7 and GSTZ1 are associated with human longevity: A two-stage case-control study using a multilocus genotyping. *Age (Dordr)* **2013**, *35*, 1357–1366. [CrossRef]

150. Montesanto, A.; Crocco, P.; Tallaro, F.; Pisani, F.; Mazzei, B.; Mari, V.; Corsonello, A.; Lattanzio, F.; Passarino, G.; Rose, G. Common polymorphisms in nitric oxide synthase (NOS) genes influence quality of aging and longevity in humans. *Biogerontology* **2013**, *14*, 177–186. [CrossRef]

151. Rose, G.; Crocco, P.; De Rango, F.; Montesanto, A.; Passarino, G. Further support to the un-coupling-to-survive theory: The genetic variation of human UCP genes is associated with longevity. *PLoS ONE* **2011**, *6*, e29650. [CrossRef]

152. Crocco, P.; Montesanto, A.; Passarino, G.; Rose, G. A common polymorphism in the UCP3 promoter influences hand grip strength in elderly people. *Biogerontology* **2011**, *12*, 265–271. [CrossRef]

153. Soerensen, M.; Thinggaard, M.; Nygaard, M.; Dato, S.; Tan, Q.; Hjelmborg, J.; Andersen-Ranberg, K.; Stevnsner, T.; Bohr, V.A.; Kimura, M.; et al. Genetic variation in TERT and TERC and human leukocyte telomere length and longevity: A cross-sectional and longitudinal analysis. *Aging Cell* **2012**, *11*, 223–227. [CrossRef]

154. Kim, S.; Bi, X.; Czarny-Ratajczak, M.; Dai, J.; Welsh, D.A.; Myers, L.; Welsch, M.A.; Cherry, K.E.; Arnold, J.; Poon, L.W.; et al. Telomere maintenance genes SIRT1 and XRCC6 impact age-related decline in telomere length but only SIRT1 is associated with human longevity. *Bi-ogerontology* **2012**, *13*, 119–131. [CrossRef]

155. Atzmon, G.; Cho, M.; Cawthon, R.M.; Budagov, T.; Katz, M.; Yang, X.; Siegel, G.; Bergman, A.; Huffman, D.M.; Schechter, C.B.; et al. Evolution in health and medicine Sackler colloquium: Genetic variation in human telomerase is associated with telomere length in Ashkenazi centenarians. *Proc. Natl. Acad. Sci. USA* **2010**, *107*, 1710–1717. [CrossRef]

156. Blackburn, E.H.; Greider, C.W.; Szostak, J.W. Telomeres and telomerase: The path from maize, Tetrahymena and yeast to human cancer and aging. *Nat. Med.* **2006**, *12*, 1133–1138. [CrossRef]

157. Joeng, K.S.; Song, E.J.; Lee, K.J.; Lee, J. Long lifespan in worms with long telomeric DNA. *Nat. Genet.* **2004**, *36*, 607–611. [CrossRef]

158. Tomas-Loba, A.; Flores, I.; Fernandez-Marcos, P.J.; Cayuela, M.L.; Maraver, A.; Tejera, A.; Borras, C.; Matheu, A.; Klatt, P.; Flores, J.M.; et al. Telomerase reverse transcriptase delays aging in cancer-resistant mice. *Cell* **2008**, *135*, 609–622. [CrossRef]

159. Kaeberlein, M.; McVey, M.; Guarente, L. The SIR2/3/4 complex and SIR2 alone promote longevity in Saccharomyces cerevisiae by two different mechanisms. *Genes Dev.* **1999**, *13*, 2570–2580. [CrossRef]

160. Lin, S.J.; Defossez, P.A.; Guarente, L. Requirement of NAD and SIR2 for life-span extension by calorie restriction in Saccharomyces cerevisiae. *Science* **2000**, *289*, 2126–2128. [CrossRef]

161. Tissenbaum, H.A.; Guarente, L. Increased dosage of a sir-2 gene extends lifespan in Caeno-rhabditis elegans. *Nature* **2001**, *410*, 227–230. [CrossRef]

162. Park, S.; Mori, R.; Shimokawa, I. Do sirtuins promote mammalian longevity? A critical review on its relevance to the longevity effect induced by calorie restriction. *Mol. Cells* **2013**, *35*, 474–480. [CrossRef]

163. Soerensen, M.; Dato, S.; Tan, Q.; Thinggaard, M.; Kleindorp, R.; Beekman, M.; Suchiman, H.E.; Jacobsen, R.; McGue, M.; Stevnsner, T.; et al. Evidence from case-control and longitudinal studies supports associations of genetic variation in APOE, CETP, and IL6 with human longevity. *Age (Dordr)* **2013**, *35*, 487–500. [CrossRef]

164. Johnson, S.C.; Rabinovitch, P.S.; Kaeberlein, M. mTOR is a key modulator of ageing and age-related disease. *Nature* **2013**, *493*, 338–345. [CrossRef]

165. Corder, E.H.; Saunders, A.M.; Strittmatter, W.J.; Schmechel, D.E.; Gaskell, P.C.; Small, G.W.; Roses, A.D.; Haines, J.L.; Pericak-Vance, M.A. Gene dose of apolipoprotein E type 4 allele and the risk of Alzheimer's disease in late onset families. *Science* **1993**, *261*, 921–923. [CrossRef] [PubMed]

166. Kervinen, K.; Savolainen, M.J.; Salokannel, J.; Hynninen, A.; Heikkinen, J.; Ehnholm, C.; Koistinen, M.J.; Kesaniemi, Y.A. Apolipoprotein E and B polymorphisms–Longevity factors assessed in nonagenarians. *Atherosclerosis* **1994**, *105*, 89–95. [CrossRef]

167. Schachter, F.; Faure-Delanef, L.; Guenot, F.; Rouger, H.; Froguel, P.; Lesueur-Ginot, L.; Cohen, D. Genetic associations with human longevity at the APOE and ACE loci. *Nat. Genet.* **1994**, *6*, 29–32. [CrossRef] [PubMed]

168. Huebbe, P.; Nebel, A.; Siegert, S.; Moehring, J.; Boesch-Saadatmandi, C.; Most, E.; Pallauf, J.; Egert, S.; Muller, M.J.; Schreiber, S.; et al. APOE epsilon4 is associated with higher vitamin D levels in targeted replacement mice and humans. *FASEB J.* **2011**, *25*, 3262–3270. [CrossRef] [PubMed]

169. Gerdes, L.U.; Jeune, B.; Ranberg, K.A.; Nybo, H.; Vaupel, J.W. Estimation of apolipoprotein E genotype-specific relative mortality risks from the distribution of genotypes in centenarians and middle-aged men: Apolipoprotein E gene is a "frailty gene," not a "longevity gene". *Genet. Epidemiol.* **2000**, *19*, 202–210. [CrossRef]

170. Vanfleteren, J.R.; Braeckman, B.P. Mechanisms of life span determination in Caenorhabditis elegans. *Neurobiol. Aging* **1999**, *20*, 487–502. [CrossRef]

171. Murakami, S.; Johnson, T.E. A genetic pathway conferring life extension and resistance to UV stress in Caenorhabditis elegans. *Genetics* **1996**, *143*, 1207–1218.

172. Tatar, M.; Kopelman, A.; Epstein, D.; Tu, M.P.; Yin, C.M.; Garofalo, R.S. A mutant Drosophila insulin receptor homolog that extends life-span and impairs neuroendocrine function. *Science* **2001**, *292*, 107–110. [CrossRef]

173. Steger, R.W.; Bartke, A.; Cecim, M. Premature ageing in transgenic mice expressing different growth hormone genes. *J. Reprod. Fertil. Suppl.* **1993**, *46*, 61–75.

174. Paolisso, G.; Gambardella, A.; Ammendola, S.; D'Amore, A.; Balbi, V.; Varricchio, M.; D'Onofrio, F. Glucose tolerance and insulin action in healthy centenarians. *Am. J. Physiol.* **1996**, *270*, E890–E894.

175. Soerensen, M.; Dato, S.; Tan, Q.; Thinggaard, M.; Kleindorp, R.; Beekman, M.; Jacobsen, R.; Suchiman, H.E.; De Craen, A.J.; Westendorp, R.G.; et al. Human longevity and variation in GH/IGF-1/insulin signaling, DNA damage signaling and repair and pro/antioxidant pathway genes: Cross sectional and longitudinal studies. *Exp. Gerontol.* **2012**, *47*, 379–387. [CrossRef]

176. Kojima, T.; Kamei, H.; Aizu, T.; Arai, Y.; Takayama, M.; Nakazawa, S.; Ebihara, Y.; Inagaki, H.; Masui, Y.; Gondo, Y.; et al. Association analysis between longevity in the Japanese population and polymorphic variants of genes involved in insulin and insulin-like growth factor 1 signaling pathways. *Exp. Gerontol.* **2004**, *39*, 1595–1598. [CrossRef]

177. Van Heemst, D.; Beekman, M.; Mooijaart, S.P.; Heijmans, B.T.; Brandt, B.W.; Zwaan, B.J.; Slagboom, P.E.; Westendorp, R.G. Reduced insulin/IGF-1 signalling and human longevity. *Aging Cell* **2005**, *4*, 79–85. [CrossRef]

178. Bonafe, M.; Barbieri, M.; Marchegiani, F.; Olivieri, F.; Ragno, E.; Giampieri, C.; Mugianesi, E.; Centurelli, M.; Franceschi, C.; Paolisso, G. Polymorphic variants of insulin-like growth factor I (IGF-I) receptor and phosphoinositide 3-kinase genes affect IGF-I plasma levels and human longevity: Cues for an evolutionarily conserved mechanism of life span control. *J. Clin. Endocrinol. Metab.* **2003**, *88*, 3299–3304. [CrossRef]

179. Stessman, J.; Maaravi, Y.; Hammerman-Rozenberg, R.; Cohen, A.; Nemanov, L.; Gritsenko, I.; Gruberman, N.; Ebstein, R.P. Candidate genes associated with ageing and life expectancy in the Jerusalem longitudinal study. *Mech. Ageing Dev.* **2005**, *126*, 333–339. [CrossRef]

180. Rose, G.; Crocco, P.; D'Aquila, P.; Montesanto, A.; Bellizzi, D.; Passarino, G. Two variants lo-cated in the upstream enhancer region of human UCP1 gene affect gene expression and are correlated with human longevity. *Exp. Gerontol.* **2011**, *46*, 897–904. [CrossRef]

181. Li, Y.; Wang, W.J.; Cao, H.; Lu, J.; Wu, C.; Hu, F.Y.; Guo, J.; Zhao, L.; Yang, F.; Zhang, Y.X.; et al. Genetic association of FOXO1A and FOXO3A with longevity trait in Han Chinese populations. *Hum. Mol. Genet.* **2009**, *18*, 4897–4904. [CrossRef]

182. Willcox, B.J.; Donlon, T.A.; He, Q.; Chen, R.; Grove, J.S.; Yano, K.; Masaki, K.H.; Willcox, D.C.; Rodriguez, B.; Curb, J.D. FOXO3A genotype is strongly associated with human longevity. *Proc. Natl. Acad. Sci. USA* **2008**, *105*, 13987–13992. [CrossRef]

183. Anselmi, C.V.; Malovini, A.; Roncarati, R.; Novelli, V.; Villa, F.; Condorelli, G.; Bellazzi, R.; Puca, A.A. Association of the FOXO3A locus with extreme longevity in a southern Italian centenar-ian study. *Rejuvenation Res.* **2009**, *12*, 95–104. [CrossRef] [PubMed]

184. Flachsbart, F.; Caliebe, A.; Kleindorp, R.; Blanche, H.; Von Eller-Eberstein, H.; Nikolaus, S.; Schreiber, S.; Nebel, A. Association of FOXO3A variation with human longevity confirmed in German centenarians. *Proc. Natl. Acad. Sci. USA* **2009**, *106*, 2700–2705. [CrossRef] [PubMed]

185. Soerensen, M.; Dato, S.; Christensen, K.; McGue, M.; Stevnsner, T.; Bohr, V.A.; Christiansen, L. Replication of an association of variation in the FOXO3A gene with human longevity using both case-control and longitudinal data. *Aging Cell* **2010**, *9*, 1010–1017. [CrossRef] [PubMed]

186. Barbieri, M.; Bonafe, M.; Franceschi, C.; Paolisso, G. Insulin/IGF-I-signaling pathway: An evo-lutionarily conserved mechanism of longevity from yeast to humans. *Am. J. Physiol. Endocrinol. Metab.* **2003**, *285*, E1064–E1071. [CrossRef] [PubMed]

187. Giannattasio, S.; Mirisola, M.G.; Mazzoni, C. Cell Stress, Metabolic Reprogramming, and Cancer. *Front. Oncol.* **2018**, *8*, 236. [CrossRef] [PubMed]

188. Kaeberlein, M.; Powers, R.W., 3rd; Steffen, K.K.; Westman, E.A.; Hu, D.; Dang, N.; Kerr, E.O.; Kirkland, K.T.; Fields, S.; Kennedy, B.K. Regulation of yeast replicative life span by TOR and Sch9 in response to nutrients. *Science* **2005**, *310*, 1193–1196. [CrossRef] [PubMed]

189. McElwee, J.J.; Schuster, E.; Blanc, E.; Piper, M.D.; Thomas, J.H.; Patel, D.S.; Selman, C.; Withers, D.J.; Thornton, J.M.; Partridge, L.; et al. Evolutionary conservation of regulated lon-gevity assurance mechanisms. *Genome Biol.* **2007**, *8*, R132. [CrossRef] [PubMed]

190. Tullet, J.M.; Hertweck, M.; An, J.H.; Baker, J.; Hwang, J.Y.; Liu, S.; Oliveira, R.P.; Baumeister, R.; Blackwell, T.K. Direct inhibition of the longevity-promoting factor SKN-1 by insulin-like signaling in C. elegans. *Cell* **2008**, *132*, 1025–1038. [CrossRef] [PubMed]

191. Taormina, G.; Mirisola, M.G. Calorie restriction in mammals and simple model organisms. *BioMed Res. Int.* **2014**, *2014*, 308690. [CrossRef] [PubMed]

192. Levine, M.E.; Suarez, J.A.; Brandhorst, S.; Balasubramanian, P.; Cheng, C.W.; Madia, F.; Fontana, L.; Mirisola, M.G.; Guevara-Aguirre, J.; Wan, J.; et al. Low protein intake is associated with a major reduction in IGF-1, cancer, and overall mortality in the 65 and younger but not older population. *Cell Metab.* **2014**, *19*, 407–417. [CrossRef] [PubMed]

GCAT
TACG
GCAT
genes

MDPI

Review

The Genetic Variability of *APOE* in Different Human Populations and Its Implications for Longevity

Paolo Abondio [1,*], Marco Sazzini [1], Paolo Garagnani [2], Alessio Boattini [1], Daniela Monti [3], Claudio Franceschi [4], Donata Luiselli [5,†] and Cristina Giuliani [1,6,*,†]

[1]   Laboratory of Molecular Anthropology & Centre for Genome Biology, Department of Biological, Geological and Environmental Sciences, University of Bologna, 40126 Bologna, Italy; marco.sazzini2@unibo.it (M.S.); alessio.boattini2@unibo.it (A.B.)
[2]   Department of Experimental, Diagnostic, and Specialty Medicine (DIMES), University of Bologna, 40126 Bologna, Italy; paolo.garagnani2@unibo.it
[3]   Department of Experimental and Clinical Biomedical Sciences "Mario Serio", University of Florence, 50134 Florence, Italy; daniela.monti@unifi.it
[4]   IRCCS Istituto delle Scienze Neurologiche di Bologna, 40139 Bologna, Italy; claudio.franceschi@unibo.it
[5]   Department of Cultural Heritage (DBC), University of Bologna, Ravenna Campus, 48121 Ravenna, Italy; donata.luiselli@unibo.it
[6]   School of Anthropology and Museum Ethnography, University of Oxford, OX2 6PE Oxford, UK
[*]   Correspondence: paolo.abondio2@unibo.it (P.A.); cristina.giuliani2@unibo.it (C.G.)
[†]   Equally contributing authors.

Received: 31 January 2019; Accepted: 12 March 2019; Published: 15 March 2019

Abstract: Human longevity is a complex phenotype resulting from the combinations of context-dependent gene-environment interactions that require analysis as a dynamic process in a cohesive ecological and evolutionary framework. Genome-wide association (GWAS) and whole-genome sequencing (WGS) studies on centenarians pointed toward the inclusion of the apolipoprotein E (*APOE*) polymorphisms ε2 and ε4, as implicated in the attainment of extreme longevity, which refers to their effect in age-related Alzheimer's disease (AD) and cardiovascular disease (CVD). In this case, the available literature on *APOE* and its involvement in longevity is described according to an anthropological and population genetics perspective. This aims to highlight the evolutionary history of this gene, how its participation in several biological pathways relates to human longevity, and which evolutionary dynamics may have shaped the distribution of *APOE* haplotypes across the globe. Its potential adaptive role will be described along with implications for the study of longevity in different human groups. This review also presents an updated overview of the worldwide distribution of *APOE* alleles based on modern day data from public databases and ancient DNA samples retrieved from literature in the attempt to understand the spatial and temporal frame in which present-day patterns of *APOE* variation evolved.

Keywords: apolipoprotein E; APOE; longevity; populations; genomics

---

1. Introduction

The study of *APOE* and its isoforms has spread in all the studies about the genetics of human longevity and this is one of the first genes that emerged in candidate-gene studies and in genome-wide analysis in different human populations. The pleiotropic roles of this gene as well as the pattern of variability across different human groups provide an interesting perspective on the analysis of the evolutionary relationship between human genetics, environmental variables, and the attainment of extreme longevity as a healthy phenotype. In the present review, the following topics will be discussed.

1.   *APOE* gene and protein structure and function, including the latest theoretical models describing its mechanism of action
2.   The role of *APOE* in human longevity, its physiological functions, and the involvement in pathological traits in modern populations
3.   *APOE* evolution and variability among human populations, including a novel analysis of modern and ancient data
4.   The evolutionary mechanisms that maintained *APOE* deleterious variants in modern human populations.

2. *APOE* Structure and Models

Human APOE is a 299-amino acid long protein (34 kDa in weight) belonging to the family of amphiphilic exchangeable apolipo-proteins that is expressed in hepatocytes, monocytes/macrophages, adipocytes, astrocytes, and kidney cells [1–4]. Structural studies have shown two independently-folded domains for the lipid-free protein: an N-terminal elongated domain (residues 1–167) forms a 4 α-helix cluster in which non-polar residues face the inside of the protein, while the C-terminal domain (residues 206–299) has a more relaxed structure, with α-helices generating a largely exposed hydrophobic surface [5,6]. These domains are connected by an unstructured hinge that provides a large degree of mobility, which is necessary for the protein to fulfill its primary function in the hepatic and extra-hepatic uptake of plasma lipoprotein and cholesterol [7].

The N-terminal domain contains the low-density-lipoprotein receptor (LDLR) binding region, which is a cluster of basic arginine and lysine residues, spanning between positions 135 and 150 in helix 4 (an Arg-172 residue in the hinge is also necessary for the binding function [8]).A stretch of hydrophobic residues at the end of the C-terminal domain (residues 260–299) is deemed to be responsible for binding the protein to lipids as well as for directing oligomerization of lipid-free ApoE. Since the monomer is the form that binds to lipids, oligomer dissociation appears to be the rate-limiting step of protein lipidation [9,10].

The gene itself is located on chromosome 19:q13.3, together with the apoC genes *APOC1*, *APOC2*, and *APOC4*, which are members of the exchangeable lipoprotein family, and in proximity to the mitochondrial translocase of the outer membrane gene (*TOMM40*). This is another locus involved in the development of AD [11–15].

As represented in Figure 1, the combination of two mutations at the *APOE* gene (rs7412 C/T and rs429358 C/T) gives rise to the three main protein variants, called ε2, ε3, and ε4 (or, alternatively, *APOE2*, *APOE3* and *APOE4*) [16–18]. Isoform ε3 has a cysteine in position 112 and an arginine residue in position 158, while isoform ε2 has two cysteine residues and isoform ε4 has two arginine residues. Several other mutations can act on this background to nuance the effects of the three main variants and are involved in diverse cardiovascular pathologies, as reported, for example, in a recent review by Matsunaga and Saito [19].

While the difference in sequence is limited to a couple of residues, this has a great impact on the protein biophysical and, consequently, functional properties, since the change in structural features of APOE provides insight on the different behavior of its isoforms [20–26].

In particular, the Arg158Cys mutation in isoform ε2 reduces the affinity of the protein for the LDLR 50-to-100-fold [27] due to the removal of a crucial electrostatic interaction with Asp154. Mutating this residue to a neutral alanine has shown that the isoform fully recovers its functionality [28].

The mutation Cys112Arg in isoform ε4 does not change its affinity for the receptor but its preference for lipoprotein binding shifts from HDL (as do ε3 and ε2) to LDL/VLDL. This occurs because charged residues that should be buried in the protein core are, instead, propelled outwards and can establish trans-domain interactions that modify the protein structure and, therefore, lipoprotein preference, possibly by hindering overall dynamics [29,30]. Mutagenesis experiments proved effective in re-establishing the preference of isoform ε4 for HDL [17,29,31,32].

Both domain interactions and intermolecular interactions have been recently confirmed by using Forster Resonance Energy Transfer assay (FRET), which is a method to quantify the exchange of energy between two fluorescent tags attached to the ends of the APOE protein. These experiments showed that there is a consistently significant difference among isoforms, with ε4 showing a higher degree of energy transfer for both domain interaction and polymerization. However, a different study asserted that conformational changes appeared to reduce the propensity of this isoform to self-stabilize in tetramers [33,34].

Denaturation experiments aimed at testing protein stability again showed different behaviors for the three isoforms, with the ε2 N-terminal domain being the most resistant and being followed by ε3 and ε4, which is the least resistant isoform, but shows a higher number of stable intermediates between its folded and unfolded forms [35–39]. This has been interpreted as isoform ε4 assuming partially unfolded stable states at different pH in basic environments, facilitating large conformational changes and, in doing so, increasing the remodeling rate of lipoprotein particles. This has also been noted with other exchangeable apolipo-proteins, such as APOAI and APOAII [38–41]. Higher ε4 catabolism, although being not an index of overall increased efficiency in plasma lipoprotein clearance, may justify why *APOE4* homozygotes have a lower plasma APOE concentration [42–45]. On the other hand, it has been suggested that partially folded APOE is more sensitive to proteolysis of the domain-connecting hinge and that isoform ε4 may be more easily flagged as "misfolded" due to domain interaction, particularly in the brain [46–50].

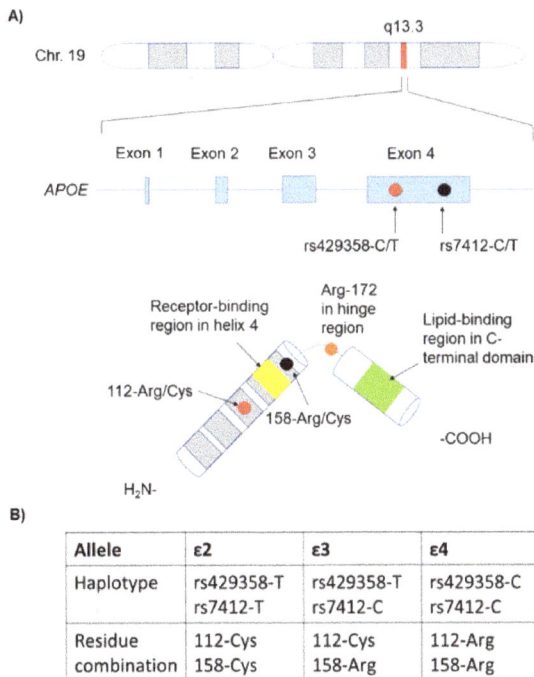

Allele	ε2	ε3	ε4
Haplotype	rs429358-T rs7412-T	rs429358-T rs7412-C	rs429358-C rs7412-C
Residue combination	112-Cys 158-Cys	112-Cys 158-Arg	112-Arg 158-Arg

Figure 1. Polymorphisms underlying the three main *APOE* variants in humans. (**A**) Chromosome location, gene structure, identity of the mutating sites in the gene, and the corresponding mutating residues in the context of the protein structure. In yellow, it is indicated as the receptor-binding region in helix 4 and, in green, it is the lipid-binding region in the C-terminal domain. Red and black dots indicate the genetic variants in *APOE* and their position in the genomic and protein sequences, respectively. (**B**) Table reporting the haplotypes and corresponding residue combination associated to each *APOE* allele.

It is also important to remember that no definitive mechanism for how APOE binds to lipids has been elucidated even though different hypotheses have emerged over the years, especially in relation to the implication of its isoforms in pathological traits. Starting from the concept of "molten globule" [36,51], a hairpin model has been proposed assuming that the protein bends itself so that the LDLR-binding motif is exposed at one extremity of the structure [31,52–55]. Other studies have suggested a conformational heterogeneity of bound apoE, observing that LDLR binding affinity, while higher in the bound protein than in the lipid-free protein, is modulated by the particle size, its lipid composition, and the presence of other bound lipoproteins [31,52,56–58].

A revised model has been recently proposed and considers the high proportion of intrinsically disordered regions in the protein (up to a third of the whole molecular structure), multiple interactions between the two domains, the presence of evolutionarily conserved residues, and structural differences that may justify the lipid-binding preferences of isoforms ε3 and ε4 [20,59]. The authors of this work also argue that most structural studies on lipid-bound apoE make use of the hepatocyte-secreted protein and plasma lipids, but that the lipid composition in the brain is different and the current models may fail to address lipidation mechanisms of astrocyte-synthesized APOE [59].

3. *APOE* Function and Pathology

Multiple lipid-related physiological functions are associated with *APOE*. In particular, isoform ε3 helps in maintaining the structural integrity of cholesterol-rich lipoproteins and enhances their solubilization in blood plasma, regulates lipid homeostasis of both hepatic and non-hepatic tissues, facilitates lipid internalization in cells and, when expressed by lipid-laden macrophages after cellular clearance, activates the reverse cholesterol transport, redirecting any excess of cholesterol to the liver for elimination [60–63].

The *APOE* genotype accounts for the vast majority of AD risk and AD pathology: inheriting one copy of *APOE4* raises a person's risk of developing the disease fourfold, while, with two copies, the risk increases 12-fold [64]. Raber and colleagues and, at the same time, Saunders and colleagues reported that clinical data regarding the association of the ε4 allele with AD suggests that 50% of AD is associated with the ε4 allele in the United States [65,66]. *APOE4* may be responsible for the accelerated formation of β-pleated amyloid, as supported by studies showing that individuals with two copies of the *APOE* ε4 allele have a higher risk and earlier onset than heterozygous subject [67]. Moreover, a significant increase in risk of EOAD (early-onset Alzheimer's disease) was found for individuals homozygous for *APOE4* regardless of family history of dementia, but an increase in EOAD risk for *APOE4* heterozygotes could only be shown in subjects with a positive family history [68].

Experiments with knock-out mice have proven that failed expression of *APOE* leads to a shortened lifespan due to the emergence of typically age-related phenotypes like an altered lipoprotein profile (the forefront of atherosclerosis and cardiovascular disease), neurological disorders, type II diabetes, deficits in immune response, and elevated markers of oxidative stress [69–75]. Moreover, the *APOE* variants determining the three isoforms ε2, ε3, and ε4 have also been associated with the modulation of body mass index (BMI) at statistical significance ($p < 10^{-3}$) in a meta-analysis including 27,863 individuals from seven longitudinal cohort studies [76]. This highlights, on one hand, that *APOE* is a pleiotropic gene that simultaneously affects multiple phenotypes, depending on the site of protein synthesis (in particular, liver and brain). On the other hand, this emphasizes that the manifestations of its impairment fit the definition of aging as a general decline in biological functions, decreased stress resistance, and elevated susceptibility to disease that leads to an increase in mortality with age [77–79].

Most of the research conducted at this point focused on isoform ε4 as the "functionally altered" form of APOE in the brain since this is one of the most consistent candidates associated with human longevity and the onset of AD, according to GWAS and whole genome sequencing studies [62,66,68,80].

The finding of unexpectedly large proportions of C-terminal APOE in β-amyloid plaques of ε4/ε4 homozygous AD subjects leads to the hypothesis that the partially folded protein is highly sensitive to proteolysis [46–50] and this prevents APOE in helping Aβ clearance, favoring instead its deposition [81]. By folding into a more helical structure, truncated ε4-165 was shown to have deleterious effects on this same process, which stresses that structural integrity is important for AD pathogenesis [82–84]. The link with Aβ has also been associated with a higher degree of lysosome leakage in neurons, primarily due to the enhanced lipid remodeling activity of isoform ε4 on the lysosomal membrane at a low pH [85,86].

Experiments on mice have highlighted how isoform ε4 can also cause behavioral deficits in the absence of amyloid accumulation and, as with AD in humans, spatial and memory impairments increase with age and are observed primarily in females [87–90]. Regarding neuronal plasticity, similar studies showed that isoform ε3 associated with VLDL clearly stimulates neurite extension in developing neurons by feeding their membrane with lipids, while isoform ε4 inhibits branching likely due to effects on microtubule stability mediated by the LDLR-protein signaling pathway. The ε4 isoform also inhibits GABAergic input in newly formed neurons [91–94].

Furthermore, this isoform has been associated with decreased cerebral glucose metabolism that occurs even decades before the cognitive impairment becomes apparent, which suggests an interaction with the mitochondrial membrane and components of the respiratory complexes III and IV at very early stages of the disease [95–100]. An interesting observation is that mitochondria and the endoplasmic reticulum (ER) are intimately connected via mitochondria-associated membranes (MAMs) and the protein miofusin-2, so that mitochondrial dysfunction may propagate to the ER and affect the secretory pathway [12,101]. If the protein is recognized as unfolded, the pathways of the unfolded protein response can activate an inflammatory process by stimulating NF-kB, which is a transcription and cytokine regulator that mediates the immune response in cell survival [102–104].

Isoform ε4 also shows a decrease in the anti-oxidative properties of APOE as a metal cation binding protein. In fact, APOE4 genotype correlates with a higher degree of lipid oxidation and presence of hydroxyl radical levels in the blood of post-mortem patients [71,105]. Macrophages overexpressing ε4 also display membrane oxidation and generate anion radicals and, as a stress response, an increase of the anti-inflammatory protein heme oxygenase 1, was observed [106].

Moreover, it has been noted that, because of the cholesterol binding property of APOE and the fact that cholesterol is the main component of the envelope of many human-infecting viruses, the different behaviors of isoforms ε3 and ε4 may, respectively, impede or ease infections. For example, extensive work in the last 20 years showed that herpes simplex virus HSV-1 is frequently found in the brain of elderly normal patients as well as AD-affected patients, and it is thought that isoform ε4 can facilitate the process of colonization and repeated activation of latent colonies through inflammation, which exacerbates neural decay at a younger age. It is also suggested that an antiviral therapy may be effective in slowing AD progression (see comprehensive reviews in References [107–109]). The hepatitis C virus, on the other hand, needs APOE for assembling and the host lipid metabolism is directly involved in the viral infection [110–115]. Lastly, an interesting set of studies tried to investigate a link between *APOE* and the modulation of HIV infection as a chronic disease, now that the affected individuals can live to older ages thanks to anti-retroviral therapy. Even though the overall results are somewhat contrasting, isoform ε4 seems to correlate in different cases with the development of HIV-associated neurocognitive disorders, impaired cognition, dyslipidaemia, premature brain aging, and increased chance of debilitating opportunistic infections [116–120] (see also a comprehensive review in Reference [121]).

However, one of the most notable associations to be examined is between *APOE* alleles and cardiovascular disease (CVD). A study carried out on nine cohorts (eight of European and one of Chinese ancestry) of middle-aged men recruited by the World Health Organization MONICA (Monitoring of Trends and Determinants in Cardiovascular Disease) Project showed how variation in the relative frequency of the ε4 allele could predict 40% to 75% of the variation in coronary heart disease (CHD) fatalities among populations and how a 0.01 increase in the frequency of this allele could increase CHD death rates by 24.5/100,000 [122]. A study on follow-up data for almost 1000 Danish and Finnish heart attack survivors similarly denoted that carrying this variant can be a prognostic element, as these subjects have an 80% increased risk of dying [123]. A similar conclusion is presented by a post-mortem study, performed at the Oslo University Hospital, on over 1500 individuals who died of natural causes. In the cohort of patients presenting a cardiovascular disease (35% of the total), there were significantly more ε4 carriers (34% against 29%) and significantly less ε2 carriers (12% against 14%) than in the rest of the group ($p < 0.05$) [124]. It has also been recently recognized that, not only APOE is associated to cardiovascular risk, but also with the level of unsaturated and saturated circulating fatty acids, so that some light is being shed on how environmental and dietary factors can mediate the association between APOE variants and adverse cardiovascular events [125].

The common APOE alleles ε2, ε3, and ε4 are located in a CpG island and the related SNPs impact on the quantity of CpG dinucleotide, which impacts the gene DNA methylation. A recent study showed that the DNA methylation profile of this genomic region differentiates AD brain if compared to that of control subjects [126]. Moreover, a recent study on lymphocytes showed that DNA methylation in the *APOE* gene is associated with age and shaped by genetic variants in the gene [114]. A different study in African Americans also suggested that DNA methylation in blood cells may be an early indicator of individuals at risk for dementia [127].

4. *APOE* and Human Longevity

Many studies have attempted to grasp the complexity of the genetics of human longevity [128–133]: recent findings suggest that alleles associated with this phenotype are population-specific and, at the same time, that the achievement of extreme longevity is modulated by mechanisms shared among populations [134–136]. One of the most relevant loci identified by many studies (if not all) is the *APOE* gene.

Candidate gene studies, genome-wide association studies (GWAS) on geographically diverse populations, and, more recently, whole-genome sequencing approaches have been aimed at uncovering the genetic variants that influence the longevity phenotype and *APOE* possibly due to its involvement in several post-reproductive pathologies, which has emerged as a strong candidate in most of them. In this section, a brief overview of the studies on human longevity conducted in relation to the three main variants of *APOE* is presented, with special attention to its isoforms ε2, ε3, and ε4 arising from the combination of two mutations (rs7412 C/T and rs429358 C/T) [16–18].

Several GWAS supported the association between *APOE* and the longevity trait. For example, a Japanese study including 743 centenarians and 822 middle-aged controls found a novel positive association between variant rs16835198-G of the gene *FNDC5* (which synthetizes a pro-hormone that is upregulated by muscular exercise) and *APOE* alleles in individuals with extreme longevity, which further highlights the polygenic nature of this trait [137]. A recent meta-analysis of GWAS examined data from 6036 individuals at least 90 years old against a control group of 3757 subjects that died between the ages of 55 and 80. A replication of known variants at *APOE* and *FOXO3* genes was obtained, but the authors also pointed out the difficulty in locating new alleles associated with survival past the age of 90, possibly because of heterogeneous genetic influences combined with the fact that rare variants are not usually picked up by GWAS [80]. A novel statistical method for evaluating genome-wide associations starting from previous knowledge of age-dependent and disease-related traits that overlap with longevity (i.e., informed GWAS, or iGWAS) was applied to reduce the background SNPs possibly associated with extreme ages and to amplify potential signals that could be difficult to pick up in small centenarian cohorts [138]. Accordingly, 92 SNPs at eight independent loci (including the *APOE/TOMM40* locus) were found to be associated with longevity at GWAS significance ($p < 10^{-8}$) and four of these were further replicated in three different validation cohorts including the *APOE/TOMM40* rs4420638 variant [138].

However, other studies failed to identify significant associations. For example, a study involving a Chinese cohort of 312 individuals with at least one long-lived parent (i.e., aged over 90) and 298 controls without a familial history of longevity found no significant correlation between *APOE* isoforms, age, and the levels of blood cholesterol (HDL-C) even though HDL-C levels themselves are significantly higher in the longevity group ($p = 0.0001$) [139]. The first study on a Brazilian cohort, including 220 individuals of at least 85 years of age and 232 controls averaging 72 years, was recently performed to investigate the association between *FOXO3*, *SOD2*, *SIRT1*, and *APOE* known variants and several phenotypes in oldest-olds. Only an association of two *FOXO3* alleles with gender and triglyceride levels was confirmed in this case and the authors suggested expansion of the number of samples in order to perform a more powerful analysis [140].

A similar pattern emerged from candidate gene studies, as some have highlighted putative associations between *APOE* and extreme lifespan, while others have not. For example, a study focused on three independent cohorts of centenarians from Italy, Spain, and Japan compared with healthy, younger controls confirmed the ε4 allele being negatively associated with extreme longevity in all three cases after adjustment for sex, while allele ε2 was positively associated with the same trait in the Japanese and Italian cohorts only. This highlighted that the ε4 variant appears to decrease the likelihood of reaching extreme ages across ethnicity and geographic origin [141]. A recently published paper on 450 individuals of Ashkenazi Jewish ancestry at least 95 years of age contrasted with 500 controls without a history of familial longevity, which undertook a full analysis of the coding and regulatory regions of *APOE*. Two common regulatory variants were, thus, found in the proximal promoter of the gene (rs405509 and rs769449), which is significantly depleted in the elderly group ($p < 0.036$). Moreover, a significant enrichment of the ε2 allele ($p = 0.003$) and the ε2/ε3 genotype ($p = 0.005$), as well as a reduction of the ε3/ε4 genotype ($p = 0.005$) were observed in the same group [142]. Two recent reviews and meta-analyses of polymorphisms associated with human longevity recovered genomic data of European and Asiatic cohorts involving centenarians (i.e., 13 cohorts [141,143–153] for the 2014 review [154], 12 cohort s [141,143,144,148,149,155–158] for the 2018 review [130]), and added newly generated data to obtain groups of at least 2700 centenarian cases and 11,000 younger controls. The first study highlighted how the likelihood of reaching extreme longevity is negatively associated with carrying the ε4 allele, the ε4/ε4, ε3/ε4, or ε2/ε4 genotypes (all $p < 0.001$), while the trait is positively associated with the ε2/ε3 genotype ($p = 0.017$) [154]. The second study ascertained the homogeneity between the European and Asian groups when accounting for ethnicity. It also confirmed a significant negative association of the ε4 allele with longevity and a positive association of the ε2 variant with the same trait (which was not supported by the 2014 meta-analysis [154]) when compared to the ε3 allele ($p < 0.0001$) [130]. In another meta-analysis, data of over 28,000 individuals born between 1880 and 1975 were collected from seven studies on population longevity and familial healthy aging, with cases ranging from 96 to 119 years and controls from 0 to 99 years. Three genetic models (i.e., standard genotypic model, additive model for the effects of the ε2 allele, grouping of genotypes containing and not containing ε4) and two definitions of longevity (i.e., age at death, age reached by less than 1% of the population) were applied. Results showed that carrying the ε2 allele, but not ε4, is associated with significantly increased odds of reaching extreme longevity, with decreased risk of death when compared to the most common genotype ε3/ε3, but modest risk reduction at the most extreme ages. The opposite is observed for ε4, which acts independently from ε2 and associates with decreased odds for extended lifespan and an increased death risk that persists even at extreme ages in all groups. Furthermore, a joint haplotype analysis of five SNPs at the *PRVL2-TOMM40-APOE-APOC1* gene cluster revealed that three haplotypes were individually associated with extreme lifespan when compared to the most common haplotype. The first one, containing ε2, was associated with a 34% increase in odds of extreme longevity ($p = 7.8 \times 10^{-7}$). The second one, containing ε4, was associated with a 50% decrease in the same odds ($p = 10^{-8}$). The last one was, instead, an uncommon haplotype containing ε3 and was associated with a 20% decrease in odds for extreme longevity ($p = 0.04$), which suggests that there are SNPs at this locus that can exert a negative effect on longevity independently from the influence of the *APOE* ε4 allele [159].

A more extensive collection of GWAS and candidate gene studies performed in the last 8 years and describing *APOE* gene variants in human longevity is reported in Supplementary Table S1.

A recently published paper about genetic variants affecting viability over generations in large cohorts applied a method for testing the variability in allele frequency across different ages, after considering individual ancestry. When applied to the Genetic Epidemiology Research in Adult health and Aging (GERA) cohort and to parents of the UK Biobank participants, few common variants significantly related to mortality at specific ages were found across the genome, all tagging the *APOE* ε4 allele and the *CHRNA3* gene. When testing for viability effects of genetic variant sets, strong signals ($p < 10^{-3}$) were found relating delayed puberty with longer parental lifespan, as well as later age of first birth with longer maternal lifespan and, lastly, cholesterol levels and risk of coronary artery disease, with a marked difference between male and female participants [160].

It is worth noting that recent data from Northern European populations [148,161] clarified that *APOE* variation is associated with the likelihood of reaching extreme longevity not because it is a 'longevity gene' that 'ensures' a long life by itself, but due to the fact that it is rather a 'frailty gene' that slightly influences mortality and, particularly, ε4 is associated with an increased risk for death that persists even beyond ages reached by less than 1% of the population [159].

5. *APOE* Evolution and Variability among Human Populations

Human APOE clusters with members of the groups APOA and APOC in the superfamily of exchangeable apolipoproteins. These are structurally and functionally distinct from the non-exchangeable apolipoproteins APOB48 and APOB100, which make up the core of the lipoprotein particles [162,163].

Phylogenetic reconstruction using apolipoprotein sequences from representative eukaryotic species has shown that an ancestral form of this protein already existed before Metazoan evolution (i.e., approximately 750 Mya) and that divergence between the exchangeable and non-exchangeable families is equally ancient [162]. Focusing on the human exchangeable superfamily, a similar analysis showed that APOE clusters specifically with APOA1, APOA4, and APOA5 (the most recently identified human apolipoprotein), are separated from the cluster including APOA2, APOC3, APOC2, and APOC1 (the oldest in the cluster). It is also noteworthy that the length of the synthetized protein increases from the oldest to the youngest gene [162]. When including the insect apolipo-protein ApoLpIII in the analysis, it was found to group by sequence similarity within the human APOE cluster, instead of being an outgroup to all human exchangeable proteins. This suggests that the divergence of exchangeable apolipo-proteins occurred at an early evolutionary stage, possibly with the advent of bilateral symmetry (i.e., approximately 650 Mya), while the origin of ApoLpIII has dated back to the emergence of flying insects (i.e., 500 Mya) [162,164,165]. Nevertheless, an extensive review of phylogenetic relationships among eukaryotic apolipo-proteins is not the purpose of this review [162].

Focusing on the investigation of human-specific apolipo-proteins characteristics, comparison of the protein sequence of human and primate APOE reveals that the non-human apolipo-protein has arginine in position 112, like human isoform ε4. This suggests that ε4 is the ancestral variant and recent analyses of Denisovan DNA (a specimen of archaic human found in 2010 in the Denisova cave, in the Altai Mountains in Siberia) also corroborate such a hypothesis [166,167]. Unfortunately, this information is not yet fully disentangled for the Neanderthal genomes. The other non-synonymous variants detected among the species do not alter the size or charge of the residues and are not located in functional domains [162]. The only fundamental difference, then, involves residue 61, where humans present an arginine, while all other primates have a threonine. The Thr61Arg substitution introduces a bulkier, positively charged residue near the equally charged Arg112, by which it is projected out of the N-terminal helix bundle. This repositioning allows for Arg61 in ε4 to be involved in domain interactions that affect the isoform structure, which makes the protein less stable, but readier in binding large, lipid-rich lipoproteins. It is, however, unclear how the mutation that originated human ε4 from an ancestral *APOE* could provide a net evolutionary advantage. Theories including the consumption of cholesterol-rich meat, the presence of pathogens in uncooked foods, and increasing brain size during human evolution have been proposed as well as random DNA photooxidation following the loss of body hair [162,168].

One of the most intriguing hypotheses for the development of longevity despite the presence of a deleterious *APOE* isoform, however, postulates a link with increased physical activity, over the evolutionary history of the genus *Homo*, that helped in counterbalancing a higher risk of cardiovascular disease [169]. Haplotype analysis revealed that the origin of isoforms ε2 and ε3 in humans can be dated back to 200,000 to 300,000 years ago [170], while the increase in physical exercise occurred much earlier in time, possibly around 1.8 Mya, when *Homo erectus* abandoned the sedentary lifestyle of the forests to become a hunter-gatherer. Long foraging distances and the ability to run for extended periods of time, to either follow prey or flee from danger, require endurance and increased levels of aerobic activity, which is related to the conversion of body fat into usable energy and is in stark contrast with the cardiovascular effects induced by the ε4/ε4 haplotype [171–173]. This likely relaxed the limitation on lifespan imposed by the deleterious allele and is in accordance with fossil dating and palaeodemographic analyses that testify an increase in the number of older individuals throughout the evolution of *H. erectus* and then *H. sapiens* [174], as well as the extension of post-reproductive lifespan in concert with the development of a hunter-gatherer lifestyle [169,175,176].

However, in modern populations, isoform ε4 is only the second-most common *APOE* variant, which shows the highest frequency in indigenous populations of Central Africa (40% in Aka Pygmies, 38% in Tutsis, 33% in Zairians, and 29% in Fon), Oceania (49% in the Hui population of New Guinea, 26% in the Mowanjum aboriginal tribe of Western Australia and in Polynesians from American Samoa) and Mexico (27% for the Huychol in Nayarit [177]). Isoform ε3, instead, shows peaks of 94% in the Alberta Hutterite people of Canada, 90% in Mexican Mayas, 88% in the Basque and Sardinian populations of Europe, and 86% in Han Chinese. As highlighted in Figure 2 and reported in Supplementary Table S2, a distinct latitudinal gradient for ε4 can be observed across Europe (5% to 10% in Spain, Portugal, Italy, and Greece, up to 16% in France, Belgium, and Germany, up to 23% in the Scandinavian peninsula, with peaks of 31% in the Saami population of Finland) and it has been also reported in China (5% to 17.5% in 19 distinct populations) [178–181]. In the context of the present review, data have been also gathered for a cohort of 134 Italian centenarians and 350 healthy, younger controls, so that 484 samples were enrolled in three Italian areas (North, Center, and South Italy) and clustered according to their place of birth. DNA samples were recovered after approval by the Ethical Committee of Sant'Orsola-Malpighi University Hospital (Bologna, Italy). As shown in Supplementary Tables S3 and S4, when individuals from both groups were separately clustered by macroareas [182,183], a definite gradient could be observed for the ε4 allele in both centenarians and controls, with frequencies of 0.125 and 0.124, respectively, in Northern Italy, 0.052 and 0.063 in Central Italy, and 0.026 and 0.039 in Southern Italy. Although sample size is relatively small in the latter group, the increase in frequency from South to North at both a regional and a continental level follows a pattern that has been already observed. For example, in Italy and in Europe, for other genes involved in lipid metabolism [182,184,185], this suggests that isoform ε3 may be selected against ε4 at lower latitudes, but this does not explain the evolutionary advantage of the single amino acidic mutation Arg112Cys provided in giving rise to the now most frequent *APOE* variant worldwide [178–180,186]. Studies on this topic report a higher structural stability and functional flexibility of isoform ε3, which can also be associated with metal binding, oxidative stress resistance, micronutrient uptake, enhanced neuronal repair following damage, and an energy-conserving phenotype [187–191] (see a comprehensive review on adaptation to dietary changes in Reference [192]). However, being more adaptive and responsive to environmental changes does not justify that all the ailments of isoform ε4 is associated with, tend to be post-reproductive. Theories have been recently introduced that several derived alleles (including those at the *APOE* gene) with a protective effect on cognition after menopause may result from late-life selection through an increase in younger kin survival. The proposal of this "grandmother effect" may explain the predominance of the ε3 allele in a trans-generational way by assessing that the extension of the post-reproductive lifespan as a healthy phenotype requires the prevention of age-related cognitive decline to increase the survival of younger kin under grandparental care. Moreover, cultural transmission through generations is known to shape the social structure of modern foraging populations, which enhances the survival probability of the individuals belonging to networks that are enriched in multi-generational sharing of knowledge [175,176,193,194].

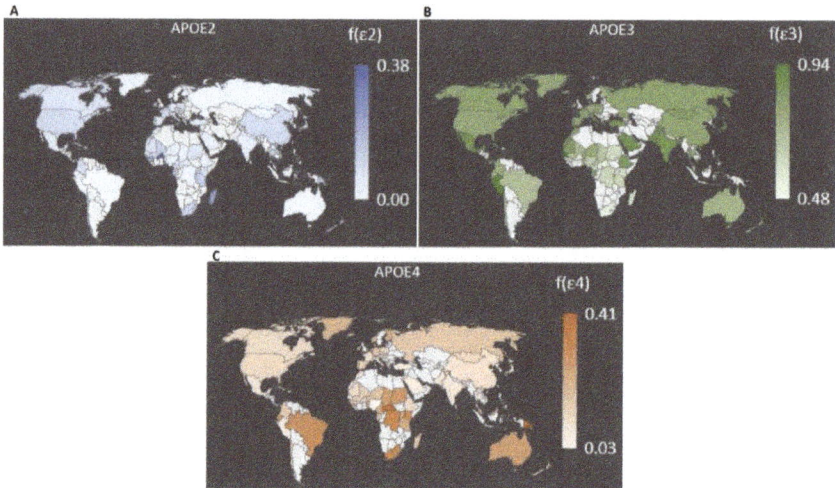

Figure 2. Frequency distribution of *APOE* alleles in 82 countries. Data from the 1000 Genome Project have been integrated with those published in Singh et al. 2006. (**A**) Frequency distribution of the ε2 variant. (**B**) Frequency distribution of the ε3 variant. (**C**) Frequency distribution of the ε4 variant.

6. *APOE* Trade-Offs

Human longevity is a complex phenotype in which small contributions from a high number of genetic variants participate to define most age-related traits in later life. Isoform ε4 of *APOE* is involved in several cardiovascular and neural pathologies that become apparent at a post-reproductive age. Many studies in the last decade tried to find explanations as to why such a deleterious variant has been maintained at high frequency in many human groups, particularly in indigenous populations of Africa and Oceania [178]. The main collected findings suggest an association between isoform ε4 and a number of population-specific and environment-related beneficial effects that compensate for the damage induced by the same variant in later life [175,176,187,189,193].

The observation that the most detrimental effects of *APOE* (CVD, AD, reduced lifespan) mainly affect individuals of affluent populations, while most African groups do not develop significant impairment despite presenting the highest frequencies of isoform ε4, prompted a study on a rural Ghanaian population characterized by high levels of mortality from widespread infectious diseases. The analyses conducted pinpointed an association between the exposure of fertile women to high pathogen levels and a higher degree of fertility (ε4 carriers have one more child than non-carriers, while ε4 homozygous women have 3.5 more children on average). Such polymorphism may be maintained because it favors reproduction in a context where limited survival at older ages spontaneously delays the detrimental effects of the isoform. Conversely, individuals living in modernized societies, less affected by pathogens and capable of reaching an older age, have no need for the positive reproductive advantage conferred by this allele and have, instead, more probability to manifest the related negative repercussions [195]. Moreover, several candidate gene studies conducted on cohorts from industrialized countries (e.g., Iran, Turkey, United States) seem to highlight a positive relationship between cardiovascular disease and thrombophilia, as induced by lipid clearance dysfunction through the ε4 variant, and occurrences of two or more consecutive miscarriages before the 20th week of gestation. These studies compared groups of affected women and fertile negative controls with at least two successful pregnancies. In all cases, a statistically significant enrichment in the ε4 variant was found for the cohorts affected by recurrent pregnancy loss as well as a significant positive association of the ε3/ε4 and ε4/ε4 genotypes with the analyzed phenotype [196–200].

Another study relating pathogen exposure to the preservation of the deleterious isoform was performed on the Tsimane population of Amazonian foragers. Results highlighted that ε4 carriers with a high eosinophil count (a sign of parasitic infection) perform better in cognitive tests than the non-infected carriers, irrespective of their age [201,202].

Some publications also support the thesis that the extremely long span of human survival beyond fertile age is an exception in the world of primates and mammals and is tightly linked to the practice of inter-generational cooperative child rearing, which potentially developed early in hunter-gatherer societies. The role of the grandmother is, in this case, equally parted in practices of active support, information transfer, and building of social networks that can result in extensive sharing of resources, which favor the survival and growth of the younger individuals. In this case, the positive effects of differential survival and reproductive success in early life are mirrored by deleterious cognitive deficiencies at an older age, when natural selection is absent [175,193].

Other studies have proposed that the main advantage provided by isoform ε3 when it first emerged, around 200,000 years ago, relates to an early shift in dietary habits. More organized hunting methods and the use of fire enhanced the quantity of fat-rich meat introduced with diet, which ultimately helped extend the human lifespan. Survival to reproductive age and beyond would, in this case, require both an efficient clearance of excess cholesterol from the blood and a stronger inflammatory response to food-borne pathogens, which is provided by the more ancient isoform ε4 [168,192].

The ε4 allele is an independent risk factor in age-related mortality and all-cause mortality. Since it hampers longevity, one would expect a general reduction of allele frequency with increasing age. However, the disease risk association seems to vary in an ethnic-related way. For example, hypertension and brain hemorrhage risks are increased only in Asian and European ε4 carriers [203,204], while African and Hispanic Americans show an increased risk for Alzheimer disease even in the absence of ε4, which allows for its accumulation in older age cohorts, because it is less detrimental [178,205]. Other studies have shown how this variant may exert negative pleiotropy, which grants protection to the infant brain and against infections at a younger age. This counterbalances the deleterious effects that may be induced later in life [206–209].

Lastly, isoform ε2 has a worldwide frequency of around 7% and a patchy distribution, with peaks in Southeast Asia, Australia, and some African populations (up to 19%) and absence in most indigenous American groups [178]. The effects of this isoform are opposite to those of ε4. Carriers show a lower risk and delayed onset of cognitive decline and a significantly reduced risk of cardiovascular disease, but increased infection rates at a young age [208,210–214]. Given the opposite effects of the two isoforms, the only current explanation for their simultaneous high frequency in several indigenous African populations is that selection acts for ε2 and ε4 against ε3, but no definitive selective mechanism has been described so far [162].

Other possible explanations for the latitudinal distribution of *APOE* variants and the maintenance of ε4 relate to its role in immunity. As highlighted by ecological and biogeographical research, there is a clear relationship between the current distribution of human infectious diseases, latitude, average temperature, humidity, and population density, with harmful bacteria flourishing in hot, wet climates and in densely populated areas of the world [215,216]. Studies involving knock-out chimeras in mice suggest that *APOE* deficiency (also mimicking reduced functionality of ε4) leads to cholesterol build-up in dendritic cell membranes, which enhances antigen presentation via lipid rafts and increasing T-cell activity (hampering macrophage function [106,217]) regardless of ensuing hypercholesterolemia. This has also been directly observed in humans, where subjects expressing the ε4 isoform have a higher activated T-cell count when compared to carriers of the other isoforms [218]. However, earlier studies on mice also highlighted that *APOE*-deficient specimens may show a significantly reduced immune response to specific pathogens by becoming more susceptible to *Lysteria monocytogenes* and *Klebsiella pneumoniae* infections [72,73,219]. As described in paragraph 3.2, several viruses require the most common form of APOE to build their particles and invade human cells. In fact, it has been observed that isoform ε4 may hamper virion synthesis and compete with the hepatitis C virus for access to LDL receptors, which reduces liver damage in exposed populations. For example, in the Italian peninsula, the North-South gradient of hepatitis C incidence overlaps with a reverse gradient in ε4 distribution [220,221]. It is less clear how the different isoforms of APOE interact with the herpes simplex virus and HIV even though the mechanisms proposed in a review by Kuhlman et al. (2010) suggest that ε4, in this case, poses less competition to cell entry, which is also helped by the enhanced presence of lipid rafts in the cell membrane [222]. A relationship between *APOE4* conservation, enhanced immune response, and pathogen distribution can be further justified by studies highlighting how carriers of this allele show higher levels of the anti-parasitic cytokine interleukin-3 (IL-3) and the pro-inflammatory tumor necrosis factor (TNFα) when exposed to endotoxins [223,224]. This seems to be especially important in the extreme case of Gram-negative infections since their toxins are membrane lipopolysaccharides (LPS) that can be collected by lipoproteins and redirected by APOE to the liver for inactivation. Reduced functionality of this protein can, thus, lead to hampered endotoxin clearance, overstimulation of macrophages, overproduction of inflammatory cytokines, and a stronger immune response leading to sepsis in the afflicted subject [225].

While local accumulation of the ε4 isoform in indigenous populations can be justified by the prevalence of infections in the absence of medical care, it can also be associated with a stronger inflammatory response to food-borne pathogens [168,192]. Other dietary factors, such as vitamin D and bone calcium assimilation, which were proven to be higher in both humans and transgenic mice carrying the ε4 allele [188,191], may have been crucial in the adaptation of populations living at extreme latitudes to the reduced amount of UV radiation. This justifies the North-South distribution of ε4 observable in Europe [188].

Many recent studies also considered a relationship between *APOE* and the gut microbiota, since, in this context, *APOE* can simultaneously exert its double role in lipid assimilation and immunity. Several experiments using *APOE* knock-out mice have shown that the diet can modulate gut microbiota composition such as with an enrichment in Firmicutes when mice were fed a typically Western diet. In turn, this relates to the amount of metabolic endotoxins in the bloodstream that stimulated a chronic inflammatory state [226,227]. On the other hand, if mice feeding on a hyperlipidic diet were immunized against their own gut microbiota, a significant decrement in serum inflammatory cytokines could be observed together with a reduction in atherosclerotic plaques, which suggests an interesting trade-off mechanism that balances the immune response against the resident microbiota with immune regulation of inflammation mediated by apolipoprotein E [228]. Other studies on obese mice and knock-out mice fed on regular chow versus a Western diet discovered that mending the loss of specific bacteria strains (e.g., *Akkermansia muciniphila*) caused by a hyperlipidic diet contrasted the enhanced permeability of the gastrointestinal tract to endotoxins and reduced vessel inflammation, fat dysmetabolism, and atherosclerosis both in normal and obese specimens [229–231]. Taking into consideration the immunomodulatory function of *APOE*, not only against bacteria, but also toward oxidized LDL found in sclerotic vessels, these observations highlight how both local and systemic responses can shape the overall arrangement of the intestinal biome [228].

Trade-off mechanisms may explain, in certain cases, issues regarding the replication of association signals for the same allele in different human populations and that several studies deem it more likely that a proportion of genetic influence on longevity (and of complex traits in general) may be explained through polygenic effects [232–234]. Furthermore, the studies performed until now did not fully address the role of rare mutations [235] nor the interaction between rare variants and *APOE* that seems to have a relevant impact on the phenotypic outcome, as supported by a recent study on the Hong Kong Chinese population [236]. Lastly, in this review, we did not address a potential limitation of trade-off mechanisms: the fact that they may be time-dependent and may be influenced by specific environmental (internal and external) conditions.

The contrast between *APOE4* and *APOE3* frequency distributions in current populations, with the former being prevalent in foraging communities and the latter being predominant in regions with relevant agricultural economy, led to the theory that the ε4 variant is a relic of a hunter-gatherer genetic background that has not adapted to the modern, energy-rich, and exercise-poor lifestyle [237]. To assess the possibility of observing the temporal scale of this transition, in the context of the present review, we built a panel of 1149 publicly available ancient genomes and selected 97 of them, with both rs7412 and rs429358 already directly genotyped (the original works including the selected samples can be found at References [238–258]). This has been done in order to avoid the introduction of bias in the dataset by imputing variants from highly deteriorated DNA, which usually presents extended regions of missing data. The samples, mapped in Figure 3 and listed in Supplementary Table S5 with details on the place of discovery and cultural context, cover the Euro-Mediterranean area and range from 1500 to 42,000 years ago. The ε3/ε3 genotype was found to be the most frequent (83%), followed by the ε4/ε4 genotype (13%), and the ε2/ε2 genotype (3%). The only heterozygote ε3/ε4 was represented by the Ust'Ishim sample, a 42,000-year old specimen of early hunter-gatherer human found in Siberia. In more detail, the ε2/ε2 individuals are Northern European samples from the Bronze Age. Despite carrying the ancestral genotype, all ε4/ε4 individuals are less than 8000 years old, with most of them being even more recent than 5000 years, while a conspicuous number of ε3/ε3 samples are much older than this, especially in the areas of Caucasus, between the Black Sea, the Caspian Sea, and the Middle East. This temporal and spatial distribution may be coherent with Palaeolithic alleles, like *APOE4*, having been reintroduced in Europe at higher frequency with the Yamnaya migration from the Steppe during the Bronze age and *APOE3* being present at higher frequencies in the Fertile Crescent prior to the Neolithic Revolution, even though both alleles were already present in the European populations as well, as highlighted by the older local specimens [238,243,245]. However, the limited number of samples available across such an extended geographic area and the chance of genotyping errors due to the highly deteriorated ancient DNA hinder the possibility of a thorough factual discussion of the results. In order to draw more elaborate conclusions, it would be useful to recover more complete and evenly distributed ancient data, both in space and in time.

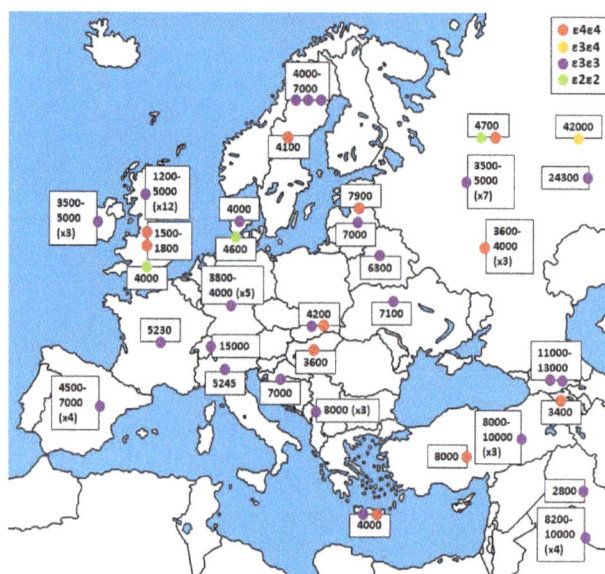

Figure 3. Distribution and approximate age of the analyzed ancient samples. Those coming from the same location and belonging to the same culture have been clustered together and share the same label. The number of grouped individuals is given in brackets.

7. Conclusions

This review reports and summarizes relevant considerations regarding *APOE* and its pivotal role in the genetics of human longevity. Both candidate-gene studies and genome-wide analysis reveal its involvement in the attainment of an extreme lifespan by exerting a pleiotropic effect in a polygenic context. In this review, some new data (on the geographic distribution of *APOE* isoforms ε2, ε3 and ε4 in centenarians and in healthy individuals from the Italian population and on public available dataset on ancient genomes) have also been considered and evaluated in the light of the most recent findings on this gene, with particular attention to the variability across human populations. In fact, the study of the variability across different human groups is crucial to understand the differences that can be observed in the association between this gene and longevity and age-related diseases. The patterns can be justified by considering the multitude of biological pathways this gene belongs to and the different environmental conditions human populations must deal with especially with regard to pathogen exposure and dietary changes. An evolutionary perspective is also crucial to understand the conservation and current worldwide distribution of *APOE* isoforms ε2, ε3, and ε4. New data regarding DNA methylation variability in different tissues will also help more clearly define the role of this gene. Moreover, the relation between population specific cultural/ecological traits and *APOE* variability (as well as other genes) are needed to disentangle the devious way from genotype to phenotype. Given the high amount of data available on this gene, we think that an evolutionary approach, such as the one proposed by evolutionary medicine [259–263], will help interpret and clarify the link between even distant (or apparently not connected) results for this gene in different populations.

Supplementary Materials: The following is available at http://www.mdpi.com/2073-4425/10/3/222/s1. Table S1: Summary of the studies published in the last eight years investigating the association between *APOE* variants and human longevity. Table S2: APOE allele frequencies in different human populations. Data used for Figure 2. Table S3: number of *APOE* alleles and haplotypes in geographically divided groups of 484 Italian centenarians and controls. Table S4: frequency of *APOE* alleles and haplotypes in geographically divided groups of 484 Italian centenarians and controls. Table S5: Summary of APOE haplotypes in ancient genomes. Data used for Figure 3.

Author Contributions: C.G., P.A., D.L., C.F., and P.G. involved in the study design. P.A., M.S., P.G., A.B., D.M., C.F., D.L., and C.G. performed the literature review. PA performed data analysis. P.A., M.S., P.G., A.B., D.M., C.F., D.L., and C.G. performed a biological interpretation. P.A. and C.G. wrote the first draft and all authors were involved in reviewing and editing. C.F. and D.M. provided data on the Italian population.

Funding: This study was supported by The European Union's Seventh Framework Program to CF (grant number 602757, HUMAN), the European Union's H2020 Project to CF and PG (grant number 634821, PROPAG-AGING), and the JPco-fuND to CF (ADAGE) supported this study.

Conflicts of Interest: The authors declare no conflict of interest.

References

1. Blue, M.L.; Williams, D.L.; Zucker, S.; Khan, S.A.; Blum, C.B. Apolipoprotein E synthesis in human kidney, adrenal gland, and liver. *Proc. Natl. Acad. Sci. USA* **1983**, *80*, 283–287. [CrossRef] [PubMed]
2. Kockx, M.; Traini, M.; Kritharides, L. Cell-specific production, secretion, and function of apolipoprotein E. *J. Mol. Med.* **2018**, *96*, 361–371. [CrossRef] [PubMed]
3. Tedla, N.; Glaros, E.N.; Brunk, U.T.; Jessup, W.; Garner, B. Heterogeneous expression of apolipoprotein-E by human macrophages. *Immunology* **2004**, *113*, 338–347. [CrossRef] [PubMed]
4. Boyles, J.K.; Pitas, R.E.; Wilson, E.; Mahley, R.W.; Taylor, J.M. Apolipoprotein E associated with astrocytic glia of the central nervous system and with nonmyelinating glia of the peripheral nervous system. *J. Clin. Investig.* **1985**, *76*, 1501–1513. [CrossRef] [PubMed]
5. Wetterau, J.R.; Aggerbeck, L.P.; Rall, S.C.; Weisgraber, K.H. Human apolipoprotein E3 in aqueous solution. I. Evidence for two structural domains. *J. Biol. Chem.* **1988**, *263*, 6240–6248. [PubMed]
6. Wilson, C.; Wardell, M.R.; Weisgraber, K.H.; Mahley, R.W.; Agard, D.A. Three-dimensional structure of the LDL receptor-binding domain of human apolipoprotein E. *Science* **1991**, *252*, 1817–1822. [CrossRef]
7. Mahley, R.W.; Innerarity, T.L.; Rall, S.C.; Weisgraber, K.H. Plasma lipoproteins: Apolipoprotein structure and function. *J. Lipid Res.* **1984**, *25*, 1277–1294.
8. Morrow, J.A.; Arnold, K.S.; Dong, J.; Balestra, M.E.; Innerarity, T.L.; Weisgraber, K.H. Effect of arginine 172 on the binding of apolipoprotein E to the low density lipoprotein receptor. *J. Biol. Chem.* **2000**, *275*, 2576–2580. [CrossRef]
9. Huang, R.Y.-C.; Garai, K.; Frieden, C.; Gross, M.L. Hydrogen/deuterium exchange and electron-transfer dissociation mass spectrometry determine the interface and dynamics of apolipoprotein E oligomerization. *Biochemistry* **2011**, *50*, 9273–9282. [CrossRef]
10. Chou, C.-Y.; Lin, Y.-L.; Huang, Y.-C.; Sheu, S.-Y.; Lin, T.-H.; Tsay, H.-J.; Chang, G.-G.; Shiao, M.-S. Structural variation in human apolipoprotein E3 and E4: Secondary structure, tertiary structure, and size distribution. *Biophys. J.* **2005**, *88*, 455–466. [CrossRef]
11. Subramanian, S.; Gottschalk, W.K.; Kim, S.Y.; Roses, A.D.; Chiba-Falek, O. The effects of PPARγ on the regulation of the *TOMM40-APOE-C1* genes cluster. *Biochim. Biophys. Acta* **2017**, *1863*, 810–816. [CrossRef]
12. Roses, A.; Sundseth, S.; Saunders, A.; Gottschalk, W.; Burns, D.; Lutz, M. Understanding the genetics of *APOE* and *TOMM40* and role of mitochondrial structure and function in clinical pharmacology of Alzheimer's disease. *Alzheimers Dement.* **2016**, *12*, 687–694. [CrossRef]
13. Cervantes, S.; Samaranch, L.; Vidal-Taboada, J.M.; Lamet, I.; Bullido, M.J.; Frank-García, A.; Coria, F.; Lleó, A.; Clarimón, J.; Lorenzo, E.; et al. Genetic variation in *APOE* cluster region and Alzheimer's disease risk. *Neurobiol. Aging* **2011**, *32*, 2107.e7–2107.e17. [CrossRef]
14. Papaioannou, I.; Simons, J.P.; Owen, J.S. Targeted in situ gene correction of dysfunctional *APOE* alleles to produce atheroprotective plasma ApoE3 protein. *Cardiol. Res. Pract.* **2012**, *2012*, 148796. [CrossRef]
15. Kulminski, A.M.; Huang, J.; Wang, J.; He, L.; Loika, Y.; Culminskaya, I. Apolipoprotein E region molecular signatures of Alzheimer's disease. *Aging Cell* **2018**, *17*, e12779. [CrossRef]
16. Weisgraber, K.H.; Rall, S.C.; Mahley, R.W. Human E apoprotein heterogeneity. Cysteine-arginine interchanges in the amino acid sequence of the apo-E isoforms. *J. Biol. Chem.* **1981**, *256*, 9077–9083.
17. Weisgraber, K.H. Apolipoprotein E distribution among human plasma lipoproteins: Role of the cysteine-arginine interchange at residue 112. *J. Lipid Res.* **1990**, *31*, 1503–1511.
18. Chetty, P.S.; Mayne, L.; Lund-Katz, S.; Englander, S.W.; Phillips, M.C. Helical structure, stability, and dynamics in human apolipoprotein E3 and E4 by hydrogen exchange and mass spectrometry. *Proc. Natl. Acad. Sci. USA* **2017**, *114*, 968–973. [CrossRef]

19. Matsunaga, A.; Saito, T. Apolipoprotein E mutations: A comparison between lipoprotein glomerulopathy and type III hyperlipoproteinemia. *Clin. Exp. Nephrol.* **2014**, *18*, 220–224. [CrossRef]

20. Frieden, C.; Garai, K. Concerning the structure of apoE: Structure of apoE. *Protein Sci.* **2013**, *22*, 1820–1825. [CrossRef]

21. Huang, Y.; Mahley, R.W. Apolipoprotein E: Structure and function in lipid metabolism, neurobiology, and Alzheimer's diseases. *Neurobiol. Dis.* **2014**, *72*, 3–12. [CrossRef]

22. Mahley, R.W.; Weisgraber, K.H.; Huang, Y. Apolipoprotein E: Structure determines function, from atherosclerosis to Alzheimer's disease to AIDS. *J. Lipid Res.* **2009**, *50*, S183–S188. [CrossRef]

23. Nguyen, D.; Dhanasekaran, P.; Nickel, M.; Mizuguchi, C.; Watanabe, M.; Saito, H.; Phillips, M.C.; Lund-Katz, S. Influence of domain stability on the properties of human apolipoprotein E3 and E4 and mouse Apolipoprotein E. *Biochemistry* **2014**, *53*, 4025–4033. [CrossRef]

24. Henry, N.; Krammer, E.-M.; Stengel, F.; Adams, Q.; Van Liefferinge, F.; Hubin, E.; Chaves, R.; Efremov, R.; Aebersold, R.; Vandenbussche, G.; et al. Lipidated apolipoprotein E4 structure and its receptor binding mechanism determined by a combined cross-linking coupled to mass spectrometry and molecular dynamics approach. *PLoS Comput. Biol.* **2018**, *14*, e1006165. [CrossRef]

25. Nguyen, D.; Dhanasekaran, P.; Nickel, M.; Nakatani, R.; Saito, H.; Phillips, M.C.; Lund-Katz, S. Molecular basis for the differences in lipid and lipoprotein binding properties of human apolipoproteins E3 and E4. *Biochemistry* **2010**, *49*, 10881–10889. [CrossRef]

26. Nguyen, D.; Dhanasekaran, P.; Phillips, M.C.; Lund-Katz, S. Molecular mechanism of apolipoprotein E binding to lipoprotein particles. *Biochemistry* **2009**, *48*, 3025–3032. [CrossRef]

27. Weisgraber, K.H.; Innerarity, T.L.; Mahley, R.W. Abnormal lipoprotein receptor-binding activity of the human E apoprotein due to cysteine-arginine interchange at a single site. *J. Biol. Chem.* **1982**, *257*, 2518–2521. [PubMed]

28. Dong, L.-M.; Parkin, S.; Trakhanov, S.D.; Rupp, B.; Simmons, T.; Arnold, K.S.; Newhouse, Y.M.; Innerarity, T.L.; Weisgraber, K.H. Novel mechanism for defective receptor binding of apolipoprotein E2 in type III hyperlipoproteinemia. *Nat. Struct. Mol. Biol.* **1996**, *3*, 718–722. [CrossRef]

29. Dong, L.-M.; Weisgraber, K.H. Human Apolipoprotein E4 Domain Interaction arginine 61 and glutamic acid 255 interact to direct the preference for very low density lipoproteins. *J. Biol. Chem.* **1996**, *271*, 19053–19057. [CrossRef] [PubMed]

30. Dong, L.M.; Wilson, C.; Wardell, M.R.; Simmons, T.; Mahley, R.W.; Weisgraber, K.H.; Agard, D.A. Human apolipoprotein E. Role of arginine 61 in mediating the lipoprotein preferences of the E3 and E4 isoforms. *J. Biol. Chem.* **1994**, *269*, 22358–22365. [PubMed]

31. Hatters, D.M.; Peters-Libeu, C.A.; Weisgraber, K.H. Engineering conformational destabilization into mouse apolipoprotein E A model for a unique property of human apolipoprotein E4. *J. Biol. Chem.* **2005**, *280*, 26477–26482. [CrossRef]

32. Hatters, D.M.; Budamagunta, M.S.; Voss, J.C.; Weisgraber, K.H. Modulation of apolipoprotein E structure by domain interaction differences in lipid-bound and lipid-free forms. *J. Biol. Chem.* **2005**, *280*, 34288–34295. [CrossRef]

33. Kara, E.; Marks, J.D.; Fan, Z.; Klickstein, J.A.; Roe, A.D.; Krogh, K.A.; Wegmann, S.; Maesako, M.; Luo, C.C.; Mylvaganam, R.; et al. Isoform- and cell type-specific structure of apolipoprotein E lipoparticles as revealed by a novel Forster resonance energy transfer assay. *J. Biol. Chem.* **2017**, *292*, 14720–14729. [CrossRef]

34. Xu, Q.; Brecht, W.J.; Weisgraber, K.H.; Mahley, R.W.; Huang, Y. Apolipoprotein E4 domain interaction occurs in living neuronal cells as determined by fluorescence resonance energy transfer. *J. Biol. Chem.* **2004**, *279*, 25511–25516. [CrossRef]

35. Morrow, J.A.; Segall, M.L.; Lund-Katz, S.; Phillips, M.C.; Knapp, M.; Rupp, B.; Weisgraber, K.H. Differences in stability among the human apolipoprotein E isoforms determined by the amino-terminal domain. *Biochemistry* **2000**, *39*, 11657–11666. [CrossRef]

36. Morrow, J.A.; Hatters, D.M.; Lu, B.; Höchtl, P.; Oberg, K.A.; Rupp, B.; Weisgraber, K.H. Apolipoprotein E4 forms a Molten Globule A potential basis for its association with disease. *J. Biol. Chem.* **2002**, *277*, 50380–50385. [CrossRef]

37. Acharya, P.; Segall, M.L.; Zaiou, M.; Morrow, J.; Weisgraber, K.H.; Phillips, M.C.; Lund-Katz, S.; Snow, J. Comparison of the stabilities and unfolding pathways of human apolipoprotein E isoforms by differential scanning calorimetry and circular dichroism. *Biochim. Biophys. Acta BBA Mol. Cell Biol. Lipids* **2002**, *1584*, 9–19. [CrossRef]

38. Bychkova, V.E.; Ptitsyn, O.B. Folding intermediates are involved in genetic diseases? *FEBS Lett.* **1995**, *359*, 6–8. [CrossRef]

39. Ptitsyn, O.B.; Bychkova, V.E.; Uversky, V.N. Kinetic and equilibrium folding intermediates. *Philos. Trans. R. Soc. Lond. B* **1995**, *348*, 35–41.

40. Gursky, O.; Atkinson, D. Thermal unfolding of human high-density apolipoprotein A-1: Implications for a lipid-free molten globular state. *Proc. Natl. Acad. Sci. USA* **1996**, *93*, 2991–2995. [CrossRef]

41. Gursky, O.; Atkinson, D. High- and low-temperature unfolding of human high-density apolipoprotein A-2. *Protein Sci. Publ. Protein Soc.* **1996**, *5*, 1874–1882. [CrossRef] [PubMed]

42. Haddy, N.; Bacquer, D.D.; Chemaly, M.M.; Maurice, M.; Ehnholm, C.; Evans, A.; Sans, S.; do Martins, M.C.; Backer, G.D.; Siest, G.; et al. The importance of plasma apolipoprotein E concentration in addition to its common polymorphism on inter-individual variation in lipid levels: Results from Apo Europe. *Eur. J. Hum. Genet.* **2002**, *10*, 841–850. [CrossRef] [PubMed]

43. Simon, R.; Girod, M.; Fonbonne, C.; Salvador, A.; Clément, Y.; Lantéri, P.; Amouyel, P.; Lambert, J.C.; Lemoine, J. Total ApoE and ApoE4 isoform assays in an Alzheimer's disease case-control study by targeted mass spectrometry (n = 669): A pilot assay for methionine-containing proteotypic peptides. *Mol. Cell. Proteom.* **2012**, *11*, 1389–1403. [CrossRef]

44. Martínez-Morillo, E.; Hansson, O.; Atagi, Y.; Bu, G.; Minthon, L.; Diamandis, E.P.; Nielsen, H.M. Total apolipoprotein E levels and specific isoform composition in cerebrospinal fluid and plasma from Alzheimer's disease patients and controls. *Acta Neuropathol.* **2014**, *127*, 633–643. [CrossRef] [PubMed]

45. Rezeli, M.; Zetterberg, H.; Blennow, K.; Brinkmalm, A.; Laurell, T.; Hansson, O.; Marko-Varga, G. Quantification of total apolipoprotein E and its specific isoforms in cerebrospinal fluid and blood in Alzheimer's disease and other neurodegenerative diseases. *EuPA Open Proteomics* **2015**, *8*, 137–143. [CrossRef]

46. Brecht, W.J.; Harris, F.M.; Chang, S.; Tesseur, I.; Yu, G.-Q.; Xu, Q.; Fish, J.D.; Wyss-Coray, T.; Buttini, M.; Mucke, L.; et al. Neuron-specific apolipoprotein E4 proteolysis is associated with increased Tau phosphorylation in brains of transgenic mice. *J. Neurosci.* **2004**, *24*, 2527–2534. [CrossRef]

47. Riddell, D.R.; Zhou, H.; Atchison, K.; Warwick, H.K.; Atkinson, P.J.; Jefferson, J.; Xu, L.; Aschmies, S.; Kirksey, Y.; Hu, Y.; et al. Impact of apolipoprotein E (ApoE) polymorphism on brain ApoE levels. *J. Neurosci.* **2008**, *28*, 11445–11453. [CrossRef] [PubMed]

48. Elliott, D.A.; Tsoi, K.; Holinkova, S.; Chan, S.L.; Kim, W.S.; Halliday, G.M.; Rye, K.-A.; Garner, B. Isoform-specific proteolysis of apolipoprotein-E in the brain. *Neurobiol. Aging* **2011**, *32*, 257–271. [CrossRef]

49. Williams II, B.; Convertino, M.; Das, J.; Dokholyan, N.V. ApoE4-specific misfolded intermediate identified by molecular dynamics simulations. *PLoS Comput. Biol.* **2015**, *11*, e1004359. [CrossRef]

50. Love, J.E.; Day, R.J.; Gause, J.W.; Brown, R.J.; Pu, X.; Theis, D.I.; Caraway, C.A.; Poon, W.W.; Rahman, A.A.; Morrison, B.E.; et al. Nuclear uptake of an amino-terminal fragment of apolipoprotein E4 promotes cell death and localizes within microglia of the Alzheimer's disease brain. *Int. J. Physiol. Pathophysiol. Pharmacol.* **2017**, *9*, 40–57. [PubMed]

51. Yeh, Y.-Q.; Liao, K.-F.; Shih, O.; Shiu, Y.-J.; Wu, W.-R.; Su, C.-J.; Lin, P.-C.; Jeng, U.-S. Probing the acid-induced packing structure changes of the molten globule domains of a protein near equilibrium unfolding. *J. Phys. Chem. Lett.* **2017**, *8*, 470–477. [CrossRef]

52. Fisher, C.A.; Narayanaswami, V.; Ryan, R.O. The lipid-associated conformation of the low density lipoprotein receptor binding domain of human apolipoprotein E. *J. Biol. Chem.* **2000**, *275*, 33601–33606. [CrossRef] [PubMed]

53. Narayanaswami, V.; Maiorano, J.N.; Dhanasekaran, P.; Ryan, R.O.; Phillips, M.C.; Lund-Katz, S.; Davidson, W.S. Helix orientation of the functional domains in apolipoprotein E in discoidal high density lipoprotein particles. *J. Biol. Chem.* **2004**, *279*, 14273–14279. [CrossRef]

54. Newhouse, Y.; Peters-Libeu, C.; Weisgraber, K.H. Crystallization and preliminary X-ray diffraction analysis of apolipoprotein E-containing lipoprotein particles. *Acta Crystallograph. Sect. F Struct. Biol. Cryst. Commun.* **2005**, *61*, 981–984. [CrossRef]

55. Drury, J.; Narayanaswami, V. Examination of lipid-bound conformation of apolipoprotein E4 by pyrene excimer fluorescence. *J. Biol. Chem.* **2005**, *280*, 14605–14610. [CrossRef]

56. Krul, E.S.; Tikkanen, M.J.; Schonfeld, G. Heterogeneity of apolipoprotein E epitope expression on human lipoproteins: Importance for apolipoprotein E function. *J. Lipid Res.* **1988**, *29*, 1309–1325.

57. Saito, H.; Dhanasekaran, P.; Baldwin, F.; Weisgraber, K.H.; Lund-Katz, S.; Phillips, M.C. Lipid binding-induced conformational change in human apolipoprotein E evidence for two lipid-bound states on spherical particles. *J. Biol. Chem.* **2001**, *276*, 40949–40954. [CrossRef] [PubMed]

58. Raussens, V.; Drury, J.; Forte, T.M.; Choy, N.; Goormaghtigh, E.; Ruysschaert, J.-M.; Narayanaswami, V. Orientation and mode of lipid-binding interaction of human apolipoprotein E C-terminal domain. *Biochem. J.* **2005**, *387*, 747–754. [CrossRef] [PubMed]

59. Frieden, C.; Wang, H.; Ho, C.M.W. A mechanism for lipid binding to apoE and the role of intrinsically disordered regions coupled to domain–domain interactions. *Proc. Natl. Acad. Sci. USA* **2017**, *114*, 6292–6297. [CrossRef] [PubMed]

60. Matsuura, F.; Wang, N.; Chen, W.; Jiang, X.-C.; Tall, A.R. HDL from CETP-deficient subjects shows enhanced ability to promote cholesterol efflux from macrophages in an apoE- and ABCG1-dependent pathway. *J. Clin. Investig.* **2006**, *116*, 1435–1442. [CrossRef] [PubMed]

61. Mahley, R.W.; Huang, Y.; Weisgraber, K.H. Putting cholesterol in its place: apoE and reverse cholesterol transport. *J. Clin. Investig.* **2006**, *116*, 1226–1229. [CrossRef]

62. Ang, L.S.; Cruz, R.P.; Hendel, A.; Granville, D.J. Apolipoprotein E, an important player in longevity and age-related diseases. *Exp. Gerontol.* **2008**, *43*, 615–622. [CrossRef]

63. Ilaria, Z.; Matteo, P.; Francesco, P.; Grazia, S.; Monica, G.; Laura, C.; Franco, B. Macrophage, but not systemic, apolipoprotein E is necessary for macrophage reverse cholesterol transport in vivo. *Arterioscler. Thromb. Vasc. Biol.* **2011**, *31*, 74–80.

64. Spinney, L. Alzheimer's disease: The forgetting gene. *Nat. News* **2014**, *510*, 26. [CrossRef]

65. Raber, J.; Huang, Y.; Ashford, J.W. ApoE genotype accounts for the vast majority of AD risk and AD pathology. *Neurobiol. Aging* **2004**, *25*, 641–650. [CrossRef] [PubMed]

66. Saunders, A.M.; Strittmatter, W.J.; Schmechel, D.; George-Hyslop, P.H.; Pericak-Vance, M.A.; Joo, S.H.; Rosi, B.L.; Gusella, J.F.; Crapper-MacLachlan, D.R.; Alberts, M.J. Association of apolipoprotein E allele ε4 with late-onset familial and sporadic Alzheimer's disease. *Neurology* **1993**, *43*, 1467–1472. [CrossRef]

67. Hyman, B.T.; Gomez-Isla, T.; West, H.; Briggs, M.; Chung, H.; Growdon, J.H.; Rebeck, G.W. Clinical and neuropathological correlates of apolipoprotein E genotype in Alzheimer's disease. Window on molecular epidemiology. *Ann. N. Y. Acad. Sci.* **1996**, *777*, 158–165. [CrossRef]

68. van Duijn, C.M.; de Knijff, P.; Cruts, M.; Wehnert, A.; Havekes, L.M.; Hofman, A.; Broeckhoven, C.V. Apolipoprotein E4 allele in a population-based study of early-onset Alzheimer's disease. *Nat. Genet.* **1994**, *7*, 74. [CrossRef]

69. Piedrahita, J.A.; Zhang, S.H.; Hagaman, J.R.; Oliver, P.M.; Maeda, N. Generation of mice carrying a mutant apolipoprotein E gene inactivated by gene targeting in embryonic stem cells. *Proc. Natl. Acad. Sci. USA* **1992**, *89*, 4471–4475. [CrossRef] [PubMed]

70. Zhang, S.H.; Reddick, R.L.; Piedrahita, J.A.; Maeda, N. Spontaneous hypercholesterolemia and arterial lesions in mice lacking apolipoprotein E. *Science* **1992**, *258*, 468–471. [CrossRef] [PubMed]

71. Hayek, T.; Oiknine, J.; Brook, J.G.; Aviram, M. Increased plasma and lipoprotein lipid peroxidation in apo E-deficient mice. *Biochem. Biophys. Res. Commun.* **1994**, *201*, 1567–1574. [CrossRef] [PubMed]

72. Roselaar, S.E.; Daugherty, A. Apolipoprotein E-deficient mice have impaired innate immune responses to *Listeria monocytogenes* in vivo. *J. Lipid Res.* **1998**, *39*, 1740–1743. [PubMed]

73. de Bont, N.; Netea, M.G.; Demacker, P.N.M.; Verschueren, I.; Kullberg, B.J.; van Dijk, K.W.; van der Meer, J.W.M.; Stalenhoef, A.F.H. Apolipoprotein E knock-out mice are highly susceptible to endotoxemia and *Klebsiella pneumoniae* infection. *J. Lipid Res.* **1999**, *40*, 680–685. [CrossRef]

74. Robertson, T.A.; Dutton, N.S.; Martins, R.N.; Taddei, K.; Papadimitriou, J.M. Comparison of astrocytic and myocytic metabolic dysregulation in apolipoprotein E deficient and human apolipoprotein E transgenic mice. *Neuroscience* **2000**, *98*, 353–359. [CrossRef]

75. Moghadasian, M.H.; Mcmanus, B.M.; Nguyen, L.B.; Shefer, S.; Nadji, M.; Godin, D.V.; Green, T.J.; Hill, J.; Yang, Y.; Scudamore, C.H.; et al. Pathophysiology of apolipoprotein E deficiency in mice: Relevance to apo E-related disorders in humans. *FASEB J.* **2001**, *15*, 2623–2630. [CrossRef] [PubMed]

76. Kulminski, A.M.; Loika, Y.; Culminskaya, I.; Huang, J.; Arbeev, K.G.; Bagley, O.; Feitosa, M.F.; Zmuda, J.M.; Christensen, K.; Yashin, A.I. Independent associations of *TOMM40* and *APOE* variants with body mass index. *Aging Cell* **2019**, *18*, e12869. [CrossRef]

77. Harman, D. Aging: Overview. *Ann. N. Y. Acad. Sci.* **2001**, *928*, 1–21. [CrossRef]

78. Vasto, S.; Candore, G.; Balistreri, C.R.; Caruso, M.; Colonna-Romano, G.; Grimaldi, M.P.; Listi, F.; Nuzzo, D.; Lio, D.; Caruso, C. Inflammatory networks in ageing, age-related diseases and longevity. *Mech. Ageing Dev.* **2007**, *128*, 83–91. [CrossRef]

79. Kregel, K.C.; Zhang, H.J. An integrated view of oxidative stress in aging: Basic mechanisms, functional effects, and pathological considerations. *Am. J. Physiol.-Regul. Integr. Comp. Physiol.* **2007**, *292*, R18–R36. [CrossRef]

80. Broer, L.; Buchman, A.S.; Deelen, J.; Evans, D.S.; Faul, J.D.; Lunetta, K.L.; Sebastiani, P.; Smith, J.A.; Smith, A.V.; Tanaka, T.; et al. GWAS of longevity in CHARGE consortium confirms *APOE* and *FOXO3* candidacy. *J. Gerontol. A. Biol. Sci. Med. Sci.* **2015**, *70*, 110–118. [CrossRef] [PubMed]

81. Iurescia, S.; Fioretti, D.; Mangialasche, F.; Rinaldi, M. The pathological cross talk between apolipoprotein E and amyloid-β peptide in Alzheimer's disease: Emerging gene-based therapeutic approaches. *J. Alzheimers Dis.* **2010**, *21*, 35–48. [CrossRef] [PubMed]

82. Dafnis, I.; Stratikos, E.; Tzinia, A.; Tsilibary, E.C.; Zannis, V.I.; Chroni, A. An apolipoprotein E4 fragment can promote intracellular accumulation of amyloid peptide β42. *J. Neurochem.* **2010**, *115*, 873–884. [CrossRef] [PubMed]

83. Argyri, L.; Dafnis, I.; Theodossiou, T.A.; Gantz, D.; Stratikos, E.; Chroni, A. Molecular basis for increased risk for late-onset Alzheimer disease due to the naturally occurring L28P mutation in apolipoprotein E4. *J. Biol. Chem.* **2014**, *289*, 12931–12945. [CrossRef]

84. Dafnis, I.; Argyri, L.; Sagnou, M.; Tzinia, A.; Tsilibary, E.C.; Stratikos, E.; Chroni, A. The ability of apolipoprotein E fragments to promote intraneuronal accumulation of amyloid β peptide 42 is both isoform and size-specific. *Sci. Rep.* **2016**, *6*, 30654. [CrossRef]

85. Ji, Z.-S.; Miranda, R.D.; Newhouse, Y.M.; Weisgraber, K.H.; Huang, Y.; Mahley, R.W. Apolipoprotein E4 potentiates amyloid β peptide-induced lysosomal leakage and apoptosis in neuronal cells. *J. Biol. Chem.* **2002**, *277*, 21821–21828. [CrossRef]

86. Ji, Z.-S.; Müllendorff, K.; Cheng, I.H.; Miranda, R.D.; Huang, Y.; Mahley, R.W. Reactivity of apolipoprotein E4 and amyloid β peptide lysosomal stability and neurodegeneration. *J. Biol. Chem.* **2006**, *281*, 2683–2692. [CrossRef]

87. Buttini, M.; Orth, M.; Bellosta, S.; Akeefe, H.; Pitas, R.E.; Wyss-Coray, T.; Mucke, L.; Mahley, R.W. Expression of human apolipoprotein E3 or E4 in the brains of ApoE−/− mice: Isoform-specific effects on neurodegeneration. *J. Neurosci.* **1999**, *19*, 4867–4880. [CrossRef] [PubMed]

88. Hartman, R.E.; Wozniak, D.F.; Nardi, A.; Olney, J.W.; Sartorius, L.; Holtzman, D.M. Behavioral phenotyping of GFAP-ApoE3 and -ApoE4 transgenic mice: ApoE4 mice show profound working memory impairments in the absence of Alzheimer's-like neuropathology. *Exp. Neurol.* **2001**, *170*, 326–344. [CrossRef] [PubMed]

89. Bour, A.; Grootendorst, J.; Vogel, E.; Kelche, C.; Dodart, J.-C.; Bales, K.; Moreau, P.-H.; Sullivan, P.M.; Mathis, C. Middle-aged human apoE4 targeted-replacement mice show retention deficits on a wide range of spatial memory tasks. *Behav. Brain Res.* **2008**, *193*, 174–182. [CrossRef] [PubMed]

90. Raber, J.; Wong, D.; Buttini, M.; Orth, M.; Bellosta, S.; Pitas, R.E.; Mahley, R.W.; Mucke, L. Isoform-specific effects of human apolipoprotein E on brain function revealed in ApoE knockout mice: Increased susceptibility of females. *Proc. Natl. Acad. Sci. USA* **1998**, *95*, 10914–10919. [CrossRef]

91. Nathan, B.P.; Bellosta, S.; Sanan, D.A.; Weisgraber, K.H.; Mahley, R.W.; Pitas, R.E. Differential effects of apolipoproteins E3 and E4 on neuronal growth in vitro. *Science* **1994**, *264*, 850–852. [CrossRef] [PubMed]

92. Nathan, B.P.; Chang, K.-C.; Bellosta, S.; Brisch, E.; Ge, N.; Mahley, R.W.; Pitas, R.E. The inhibitory effect of apolipoprotein E4 on neurite outgrowth is associated with microtubule depolymerization. *J. Biol. Chem.* **1995**, *270*, 19791–19799. [CrossRef] [PubMed]

93. Holtzman, D.M.; Pitas, R.E.; Kilbridge, J.; Nathan, B.; Mahley, R.W.; Bu, G.; Schwartz, A.L. Low density lipoprotein receptor-related protein mediates apolipoprotein E-dependent neurite outgrowth in a central nervous system-derived neuronal cell line. *Proc. Natl. Acad. Sci. USA* **1995**, *92*, 9480–9484. [CrossRef]

94. Li, G.; Bien-Ly, N.; Andrews-Zwilling, Y.; Xu, Q.; Bernardo, A.; Ring, K.; Halabisky, B.; Deng, C.; Mahley, R.W.; Huang, Y. GABAergic interneuron dysfunction impairs hippocampal neurogenesis in adult apolipoprotein E4 knockin mice. *Cell Stem Cell* **2009**, *5*, 634–645. [CrossRef] [PubMed]

95. Reiman, E.M.; Chen, K.; Alexander, G.E.; Caselli, R.J.; Bandy, D.; Osborne, D.; Saunders, A.M.; Hardy, J. Correlations between apolipoprotein E ε4 gene dose and brain-imaging measurements of regional hypometabolism. *Proc. Natl. Acad. Sci. USA* **2005**, *102*, 8299–8302. [CrossRef]

96. Chang, S.; Ran Ma, T.; Miranda, R.D.; Balestra, M.E.; Mahley, R.W.; Huang, Y. Lipid- and receptor-binding regions of apolipoprotein E4 fragments act in concert to cause mitochondrial dysfunction and neurotoxicity. *Proc. Natl. Acad. Sci. USA* **2005**, *102*, 18694–18699. [CrossRef]

97. Scarmeas, N.; Habeck, C.; Hilton, J.; Anderson, K.; Flynn, J.; Park, A.; Stern, Y. APOE related alterations in cerebral activation even at college age. *J. Neurol. Neurosurg. Psychiatry* **2005**, *76*, 1440–1444. [CrossRef]

98. Nakamura, T.; Watanabe, A.; Fujino, T.; Hosono, T.; Michikawa, M. Apolipoprotein E4 (1–272) fragment is associated with mitochondrial proteins and affects mitochondrial function in neuronal cells. *Mol. Neurodegener.* **2009**, *4*, 35. [CrossRef]

99. Tanaka, M.; Vedhachalam, C.; Sakamoto, T.; Dhanasekaran, P.; Phillips, M.C.; Lund-Katz, S.; Saito, H. Effect of carboxyl-terminal truncation on structure and lipid interaction of human apolipoprotein E4. *Biochemistry* **2006**, *45*, 4240–4247. [CrossRef]

100. Chou, C.-Y.; Jen, W.-P.; Hsieh, Y.-H.; Shiao, M.-S.; Chang, G.-G. Structural and functional variations in human apolipoprotein E3 and E4. *J. Biol. Chem.* **2006**, *281*, 13333–13344. [CrossRef]

101. Tambini, M.D.; Pera, M.; Kanter, E.; Yang, H.; Guardia-Laguarta, C.; Holtzman, D.; Sulzer, D.; Area-Gomez, E.; Schon, E.A. ApoE4 upregulates the activity of mitochondria-associated ER membranes. *EMBO Rep.* **2016**, *17*, 27–36. [CrossRef] [PubMed]

102. Bravo, R.; Gutierrez, T.; Paredes, F.; Gatica, D.; Rodriguez, A.E.; Pedrozo, Z.; Chiong, M.; Parra, V.; Quest, A.F.G.; Rothermel, B.A.; et al. Endoplasmic reticulum: ER stress regulates mitochondrial bioenergetics. *Int. J. Biochem. Cell Biol.* **2012**, *44*, 16–20. [CrossRef] [PubMed]

103. Cao, S.S.; Kaufman, R.J. Endoplasmic reticulum stress and oxidative stress in cell fate decision and human disease. *Antioxid. Redox Signal.* **2014**, *21*, 396–413. [CrossRef] [PubMed]

104. Chaudhari, N.; Talwar, P.; Parimisetty, A.; Lefebvre d'Hellencourt, C.; Ravanan, P. A Molecular web: Endoplasmic reticulum stress, inflammation, and oxidative stress. *Front. Cell. Neurosci.* **2014**, *8*. [CrossRef]

105. Miyata, M.; Smith, J.D. Apolipoprotein E allele–specific antioxidant activity and effects on cytotoxicity by oxidative insults and β–amyloid peptides. *Nat. Genet.* **1996**, *14*, 55–61. [CrossRef] [PubMed]

106. Jofre-Monseny, L.; Loboda, A.; Wagner, A.E.; Huebbe, P.; Boesch-Saadatmandi, C.; Jozkowicz, A.; Minihane, A.-M.; Dulak, J.; Rimbach, G. Effects of apoE genotype on macrophage inflammation and heme oxygenase-1 expression. *Biochem. Biophys. Res. Commun.* **2007**, *357*, 319–324. [CrossRef]

107. Itzhaki, R.F. Herpes and Alzheimer's disease: Subversion in the central nervous system and how it might be halted. *J. Alzheimers Dis. JAD* **2016**, *54*, 1273–1281. [CrossRef]

108. Itzhaki, R.F. Herpes simplex virus type 1 and Alzheimer's disease: Possible mechanisms and signposts. *FASEB J.* **2017**, *31*, 3216–3226. [CrossRef] [PubMed]

109. Itzhaki, R.F. Corroboration of a major role for Herpes simplex virus type 1 in Alzheimer's disease. *Front. Aging Neurosci.* **2018**, *10*, 324. [CrossRef] [PubMed]

110. Cun, W.; Jiang, J.; Luo, G. The C-terminal α-helix domain of apolipoprotein E Is required for interaction with nonstructural protein 5A and assembly of Hepatitis C virus. *J. Virol.* **2010**, *84*, 11532–11541. [CrossRef] [PubMed]

111. Chiba-Falek, O.; Linnertz, C.; Guyton, J.; Gardner, S.D.; Roses, A.D.; McCarthy, J.J.; Patel, K. Pleiotropy and allelic heterogeneity in the *TOMM40-APOE* genomic region related to clinical and metabolic features of hepatitis C infection. *Hum. Genet.* **2012**, *131*, 1911–1920. [CrossRef]

112. Bankwitz, D.; Doepke, M.; Hueging, K.; Weller, R.; Bruening, J.; Behrendt, P.; Lee, J.-Y.; Vondran, F.W.R.; Manns, M.P.; Bartenschlager, R.; et al. Maturation of secreted HCV particles by incorporation of secreted ApoE protects from antibodies by enhancing infectivity. *J. Hepatol.* **2017**, *67*, 480–489. [CrossRef]

113. Weller, R.; Hueging, K.; Brown, R.J.P.; Todt, D.; Joecks, S.; Vondran, F.W.R.; Pietschmann, T. Hepatitis C virus strain-dependent usage of apolipoprotein E modulates assembly efficiency and specific infectivity of secreted virions. *J. Virol.* **2017**, *91*. [CrossRef]

114. Gondar, V.; Molina-Jiménez, F.; Hishiki, T.; García-Buey, L.; Koutsoudakis, G.; Shimotohno, K.; Benedicto, I.; Majano, P.L. Apolipoprotein E, but not apolipoprotein B, Is essential for efficient cell-to-cell transmission of Hepatitis C virus. *J. Virol.* **2015**, *89*, 9962–9973. [CrossRef]

115. Popescu, C.-I.; Dubuisson, J. Role of lipid metabolism in hepatitis C virus assembly and entry. *Biol. Cell* **2010**, *102*, 63–74. [CrossRef] [PubMed]

116. Chang, L.; Andres, M.; Sadino, J.; Jiang, C.S.; Nakama, H.; Miller, E.; Ernst, T. Impact of apolipoprotein E ε4 and HIV on cognition and brain atrophy: antagonistic pleiotropy and premature brain aging. *NeuroImage* **2011**, *58*, 1017–1027. [CrossRef]

117. Chang, L.; Jiang, C.; Cunningham, E.; Buchthal, S.; Douet, V.; Andres, M.; Ernst, T. Effects of APOE ε4, age, and HIV on glial metabolites and cognitive deficits. *Neurology* **2014**, *82*, 2213–2222. [CrossRef]

118. Wendelken, L.A.; Jahanshad, N.; Rosen, H.J.; Busovaca, E.; Allen, I.; Coppola, G.; Adams, C.; Rankin, K.P.; Milanini, B.; Clifford, K.; et al. ApoE ε4 is associated with cognition, brain integrity and atrophy in HIV over age 60. *J. Acquir. Immune Defic. Syndr. 1999* **2016**, *73*, 426–432. [CrossRef]

119. Suwalak, T.; Srisawasdi, P.; Puangpetch, A.; Santon, S.; Koomdee, N.; Chamnanphon, M.; Charoenyingwattana, A.; Chantratita, W.; Sukasem, C. Polymorphisms of the *ApoE* (Apolipoprotein E) gene and their influence on dyslipidemia in hiv-1-infected individuals. *Jpn. J. Infect. Dis.* **2015**, *68*, 5–12. [CrossRef]

120. Cooley, S.A.; Paul, R.H.; Fennema-Notestine, C.; Morgan, E.E.; Vaida, F.; Deng, Q.; Chen, J.A.; Letendre, S.; Ellis, R.; Clifford, D.B.; et al. Apolipoprotein E ε4 genotype status is not associated with neuroimaging outcomes in a large cohort of HIV+ individuals. *J. Neurovirol.* **2016**, *22*, 607–614. [CrossRef] [PubMed]

121. Geffin, R.; McCarthy, M. Aging and apolipoprotein E in HIV infection. *J. Neurovirol.* **2018**, *24*, 529–548. [CrossRef] [PubMed]

122. Stengard, J.H.; Weiss, K.M.; Sing, C.F. An ecological study of association between coronary heart disease mortality rates in men and the relative frequencies of common allelic variations in the gene coding for apolipoprotein E. *Hum. Genet.* **1998**, *103*, 234–241. [CrossRef]

123. Gerdes, L.U.; Gerdes, C.; Kervinen, K.; Savolainen, M.; Klausen, I.C.; Hansen, P.S.; Kesäniemi, Y.A.; Faergeman, O. The apolipoprotein ε4 allele determines prognosis and the effect on prognosis of simvastatin in survivors of myocardial infarction: A substudy of the Scandinavian simvastatin survival study. *Circulation* **2000**, *101*, 1366–1371. [CrossRef] [PubMed]

124. Kumar, N.T.; Liestøl, K.; Løberg, E.M.; Reims, H.M.; Brorson, S.-H.; Mæhlen, J. The apolipoprotein E polymorphism and cardiovascular diseases—An autopsy study. *Cardiovasc. Pathol.* **2012**, *21*, 461–469. [CrossRef] [PubMed]

125. Satizabal Claudia, L.; Samieri, C.; Davis-Plourde, K.L.; Voetsch, B.; Aparicio, H.J.; Pase, M.P.; Romero, J.R.; Helmer, C.; Vasan, R.S.; Kase, C.S.; et al. APOE and the association of fatty acids with the risk of stroke, coronary heart disease, and mortality. *Stroke* **2018**, *49*, 2822–2829. [CrossRef] [PubMed]

126. Foraker, J.; Millard, S.P.; Leong, L.; Thomson, Z.; Chen, S.; Keene, C.D.; Bekris, L.M.; Yu, C.-E. The *APOE* gene is differentially methylated in Alzheimer's disease. *J. Alzheimers Dis.* **2015**, *48*, 745–755. [CrossRef]

127. Liu, J.; Zhao, W.; Ware, E.B.; Turner, S.T.; Mosley, T.H.; Smith, J.A. DNA methylation in the *APOE* genomic region is associated with cognitive function in African Americans. *BMC Med. Genomics* **2018**, *11*. [CrossRef]

128. Giuliani, C.; Sazzini, M.; Pirazzini, C.; Bacalini, M.G.; Marasco, E.; Ruscone, G.A.G.; Fang, F.; Sarno, S.; Gentilini, D.; Di Blasio, A.M.; et al. Impact of demography and population dynamics on the genetic architecture of human longevity. *Aging* **2018**, *10*, 1947–1963. [CrossRef]

129. Sebastiani, P.; Solovieff, N.; DeWan, A.T.; Walsh, K.M.; Puca, A.; Hartley, S.W.; Melista, E.; Andersen, S.; Dworkis, D.A.; Wilk, J.B.; et al. Genetic Signatures of exceptional longevity in humans. *PLoS ONE* **2012**, *7*. [CrossRef]

130. Revelas, M.; Thalamuthu, A.; Oldmeadow, C.; Evans, T.-J.; Armstrong, N.J.; Kwok, J.B.; Brodaty, H.; Schofield, P.R.; Scott, R.J.; Sachdev, P.S.; et al. Review and meta-analysis of genetic polymorphisms associated with exceptional human longevity. *Mech. Ageing Dev.* **2018**, *175*, 24–34. [CrossRef]

131. Pilling, L.C.; Atkins, J.L.; Bowman, K.; Jones, S.E.; Tyrrell, J.; Beaumont, R.N.; Ruth, K.S.; Tuke, M.A.; Yaghootkar, H.; Wood, A.R.; et al. Human longevity is influenced by many genetic variants: Evidence from 75,000 UK Biobank participants. *Aging* **2016**, *8*, 547–560. [CrossRef]

132. Pilling, L.C.; Kuo, C.-L.; Sicinski, K.; Tamosauskaite, J.; Kuchel, G.A.; Harries, L.W.; Herd, P.; Wallace, R.; Ferrucci, L.; Melzer, D. Human longevity: 25 genetic loci associated in 389,166 UK biobank participants. *Aging* **2017**, *9*, 2504–2520. [CrossRef]

133. Newman, A.B.; Walter, S.; Lunetta, K.L.; Garcia, M.E.; Slagboom, P.E.; Christensen, K.; Arnold, A.M.; Aspelund, T.; Aulchenko, Y.S.; Benjamin, E.J.; et al. A meta-analysis of four genome-wide association studies of survival to age 90 Years or older: The cohorts for heart and aging research in Genomic Epidemiology Consortium. *J. Gerontol. A Biol. Sci. Med. Sci.* **2010**, *65A*, 478–487. [CrossRef] [PubMed]

134. De Benedictis, G.; Franceschi, C. The unusual genetics of human longevity. *Sci. Aging Knowl. Environ.* **2006**, *2006*, pe20. [CrossRef]

135. Zeng, Y.; Nie, C.; Min, J.; Liu, X.; Li, M.; Chen, H.; Xu, H.; Wang, M.; Ni, T.; Li, Y.; et al. Novel loci and pathways significantly associated with longevity. *Sci. Rep.* **2016**, *6*, 21243. [CrossRef] [PubMed]

136. Giuliani, C.; Garagnani, P.; Franceschi, C. Genetics of Human longevity within an eco-evolutionary nature-nurture framework. *Circ. Res.* **2018**, *123*, 745–772. [CrossRef] [PubMed]

137. Fuku, N.; Díaz-Peña, R.; Arai, Y.; Abe, Y.; Zempo, H.; Naito, H.; Murakami, H.; Miyachi, M.; Spuch, C.; Serra-Rexach, J.A.; et al. Epistasis, physical capacity-related genes and exceptional longevity: *FNDC5* gene interactions with candidate genes *FOXOA3* and *APOE*. *BMC Genomics* **2017**, *18*, 803. [CrossRef]

138. Fortney, K.; Dobriban, E.; Garagnani, P.; Pirazzini, C.; Monti, D.; Mari, D.; Atzmon, G.; Barzilai, N.; Franceschi, C.; Owen, A.B.; et al. Genome-wide scan informed by age-related disease identifies loci for exceptional human longevity. *PLoS Genet.* **2015**, *11*. [CrossRef] [PubMed]

139. Wang, J.; Shi, L.; Zou, Y.; Tang, J.; Cai, J.; Wei, Y.; Qin, J.; Zhang, Z. Positive association of familial longevity with the moderate-high HDL-C concentration in Bama aging Study. *Aging* **2018**, *10*, 3528–3540. [CrossRef]

140. Silva-Sena, G.G.; Camporez, D.; dos Santos, L.R.; da Silva, A.S.; Sagrillo Pimassoni, L.H.; Tieppo, A.; do Pimentel Batitucci, M.C.; Morelato, R.L.; de Paula, F. An association study of *FOXO3* variant and longevity. *Genet. Mol. Biol.* **2018**, *41*, 386–396. [CrossRef]

141. Garatachea, N.; Emanuele, E.; Calero, M.; Fuku, N.; Arai, Y.; Abe, Y.; Murakami, H.; Miyachi, M.; Yvert, T.; Verde, Z.; et al. *ApoE* gene and exceptional longevity: Insights from three independent cohorts. *Exp. Gerontol.* **2014**, *53*, 16–23. [CrossRef]

142. Ryu, S.; Atzmon, G.; Barzilai, N.; Raghavachari, N.; Suh, Y. Genetic landscape of *APOE* in human longevity revealed by high-throughput sequencing. *Mech. Ageing Dev.* **2016**, *155*, 7–9. [CrossRef]

143. Louhija, J.; Miettinen, H.E.; Kontula, K.; Tikkanen, M.J.; Miettinen, T.A.; Tilvis, R.S. Aging and genetic variation of plasma apolipoproteins. Relative loss of the apolipoprotein E4 phenotype in centenarians. *Arterioscler. Thromb. J. Vasc. Biol.* **1994**, *14*, 1084–1089. [CrossRef]

144. Schächter, F.; Faure-Delanef, L.; Guénot, F.; Rouger, H.; Froguel, P.; Lesueur-Ginot, L.; Cohen, D. Genetic associations with human longevity at the APOE and ACE loci. *Nat. Genet.* **1994**, *6*, 29–32. [CrossRef] [PubMed]

145. Asada, T.; Kariya, T.; Yamagata, Z.; Kinoshita, T.; Asaka, A. Apolipoprotein E allele in centenarians. *Neurology* **1996**, *46*, 1484. [CrossRef] [PubMed]

146. Yamagata, Z.; Asada, T.; Kinoshita, A.; Zhang, Y.; Asaka, A. Distribution of apolipoprotein E gene polymorphisms in Japanese patients with Alzheimer's disease and in Japanese centenarians. *Hum. Hered.* **1997**, *47*, 22–26. [CrossRef] [PubMed]

147. Castro, E.; Ogburn, C.E.; Hunt, K.E.; Tilvis, R.; Louhija, J.; Penttinen, R.; Erkkola, R.; Panduro, A.; Riestra, R.; Piussan, C.; et al. Polymorphisms at the Werner locus: I. Newly identified polymorphisms, ethnic variability of 1367Cy/Arg, and its stability in a population of Finnish centenarians. *Am. J. Med. Genet.* **1999**, *82*, 399–403. [CrossRef]

148. Gerdes, L.U.; Jeune, B.; Ranberg, K.A.; Nybo, H.; Vaupel, J.W. Estimation of apolipoprotein E genotype-specific relative mortality risks from the distribution of genotypes in centenarians and middle-aged men: Apolipoprotein E gene is a "frailty gene," not a "longevity gene." *Genet. Epidemiol.* **2000**, *19*, 202–210. [CrossRef]

149. Blanché, H.; Cabanne, L.; Sahbatou, M.; Thomas, G. A study of French centenarians: Are ACE and APOE associated with longevity? *Comptes Rendus Académie Sci. Ser. III Sci. Vie* **2001**, *324*, 129–135. [CrossRef]

150. Arai, Y.; Hirose, N.; Nakazawa, S.; Yamamura, K.; Shimizu, K.; Takayama, M.; Ebihara, Y.; Osono, Y.; Homma, S. Lipoprotein metabolism in Japanese centenarians: Effects of apolipoprotein E polymorphism and nutritional status. *J. Am. Geriatr. Soc.* **2001**, *49*, 1434–1441. [CrossRef] [PubMed]

151. Choi, Y.-H.; Kim, J.-H.; Kim, D.K.; Kim, J.-W.; Kim, D.-K.; Lee, M.S.; Kim, C.H.; Park, S.C. Distributions of ACE and APOE polymorphisms and their relations with dementia status in Korean centenarians. *J. Gerontol. Ser. A* **2003**, *58*, M227–M231. [CrossRef]

152. Panza, F.; Solfrizzi, V.; Colacicco, A.M.; Basile, A.M.; D'Introno, A.; Capurso, C.; Sabba, M.; Capurso, S.; Capurso, A. Apolipoprotein E (APOE) polymorphism influences serum APOE levels in Alzheimer's disease patients and centenarians. *NeuroReport* **2003**, *14*, 605. [CrossRef]

153. Capurso, C.; Solfrizzi, V.; D'Introno, A.; Colacicco, A.M.; Capurso, S.A.; Semeraro, C.; Capurso, A.; Panza, F. Interleukin 6−174 G/C promoter gene polymorphism in centenarians: No evidence of association with human longevity or interaction with apolipoprotein E alleles. *Exp. Gerontol.* **2004**, *39*, 1109–1114. [CrossRef]

154. Garatachea, N.; Marín, P.J.; Santos-Lozano, A.; Sanchis-Gomar, F.; Emanuele, E.; Lucia, A. The *ApoE* gene is related with exceptional longevity: A systematic review and meta-analysis. *Rejuvenation Res.* **2014**, *18*, 3–13. [CrossRef]

155. Rea, I.M.; Mc Dowell, I.; McMaster, D.; Smye, M.; Stout, R.; Evans, A. Apolipoprotein E alleles in nonagenarian subjects in the Belfast Elderly Longitudinal Free-living Ageing Study (BELFAST). *Mech. Ageing Dev.* **2001**, *122*, 1367–1372. [CrossRef]

156. Zubenko, G.S.; Stiffler, J.S.; Hughes, H.B.; Fatigati, M.J.; Zubenko, W.N. Genome survey for loci that influence successful aging: sample characterization, method validation, and initial results for the Y chromosome. *Am. J. Geriatr. Psychiatry* **2002**, *10*, 619–630. [CrossRef]

157. Geesaman, B.J.; Benson, E.; Brewster, S.J.; Kunkel, L.M.; Blanché, H.; Thomas, G.; Perls, T.T.; Daly, M.J.; Puca, A.A. Haplotype-based identification of a microsomal transfer protein marker associated with the human lifespan. *Proc. Natl. Acad. Sci. USA* **2003**, *100*, 14115–14120. [CrossRef] [PubMed]

158. Feng, J.; Xiang, L.; Wan, G.; Qi, K.; Sun, L.; Huang, Z.; Zheng, C.; Lv, Z.; Hu, C.; Yang, Z. Is APOE ε3 a favourable factor for the longevity: An association study in Chinese population. *J. Genet.* **2011**, *90*, 343–347. [CrossRef] [PubMed]

159. Sebastiani, P.; Gurinovich, A.; Nygaard, M.; Sasaki, T.; Sweigart, B.; Bae, H.; Andersen, S.L.; Villa, F.; Atzmon, G.; Christensen, K.; et al. APOE alleles and extreme human longevity. *J. Gerontol. Ser. A* **2019**, *74*, 44–51. [CrossRef]

160. Mostafavi, H.; Berisa, T.; Day, F.R.; Perry, J.R.B.; Przeworski, M.; Pickrell, J.K. Identifying genetic variants that affect viability in large cohorts. *PLoS Biol.* **2017**, *15*, e2002458. [CrossRef] [PubMed]

161. Christensen, K.; Johnson, T.E.; Vaupel, J.W. The quest for genetic determinants of human longevity: Challenges and insights. *Nat. Rev. Genet.* **2006**, *7*, 436–448. [CrossRef]

162. Huebbe, P.; Rimbach, G. Evolution of human apolipoprotein E (APOE) isoforms: Gene structure, protein function and interaction with dietary factors. *Ageing Res. Rev.* **2017**, *37*, 146–161. [CrossRef]

163. Luo, C.-C.; Li, W.-H.; Moore, M.N.; Chan, L. Structure and evolution of the apolipoprotein multigene family. *J. Mol. Biol.* **1986**, *187*, 325–340. [CrossRef]

164. Smith, A.F.; Owen, L.M.; Strobel, L.M.; Chen, H.; Kanost, M.R.; Hanneman, E.; Wells, M.A. Exchangeable apolipoproteins of insects share a common structural motif. *J. Lipid Res.* **1994**, *35*, 1976–1984.

165. Peterson, K.J.; Lyons, J.B.; Nowak, K.S.; Takacs, C.M.; Wargo, M.J.; McPeek, M.A. Estimating metazoan divergence times with a molecular clock. *Proc. Natl. Acad. Sci. USA* **2004**, *101*, 6536–6541. [CrossRef]

166. Reich, D.; Green, R.E.; Kircher, M.; Krause, J.; Patterson, N.; Durand, E.Y.; Viola, B.; Briggs, A.W.; Stenzel, U.; Johnson, P.L.F.; et al. Genetic history of an archaic hominin group from Denisova Cave in Siberia. *Nature* **2010**, *468*, 1053–1060. [CrossRef]

167. McIntosh, A.M.; Bennett, C.; Dickson, D.; Anestis, S.F.; Watts, D.P.; Webster, T.H.; Fontenot, M.B.; Bradley, B.J. The apolipoprotein E (*APOE*) gene appears functionally monomorphic in chimpanzees (*Pan troglodytes*). *PLoS ONE* **2012**, *7*. [CrossRef]

168. Finch, C.E.; Stanford, C.B. Meat-Adaptive genes and the evolution of slower aging in humans. *Q. Rev. Biol.* **2004**, *79*, 3–50. [CrossRef]

169. Raichlen, D.A.; Alexander, G.E. Exercise, APOE genotype, and the evolution of the human lifespan. *Trends Neurosci.* **2014**, *37*, 247–255. [CrossRef]

170. Fullerton, S.M.; Clark, A.G.; Weiss, K.M.; Nickerson, D.A.; Taylor, S.L.; Stengård, J.H.; Salomaa, V.; Vartiainen, E.; Perola, M.; Boerwinkle, E.; et al. Apolipoprotein E variation at the sequence haplotype level: Implications for the origin and maintenance of a major human polymorphism. *Am. J. Hum. Genet.* **2000**, *67*, 881–900. [CrossRef]

171. Antón, S.C.; Leonard, W.R.; Robertson, M.L. An ecomorphological model of the initial hominid dispersal from Africa. *J. Hum. Evol.* **2002**, *43*, 773–785. [CrossRef]

172. Bramble, D.M.; Lieberman, D.E. Endurance running and the evolution of *Homo*. *Nature* **2004**, *432*, 345–352. [CrossRef]

173. Malina, R.M.; Little, B.B. Physical activity: The present in the context of the past. *Am. J. Hum. Biol. Off. J. Hum. Biol. Counc.* **2008**, *20*, 373–391. [CrossRef]

174. Caspari, R.; Lee, S.-H. Older age becomes common late in human evolution. *Proc. Natl. Acad. Sci. USA* **2004**, *101*, 10895–10900. [CrossRef]

175. Hawkes, K.; O'Connell, J.F.; Jones, N.G.B.; Alvarez, H.; Charnov, E.L. Grandmothering, menopause, and the evolution of human life histories. *Proc. Natl. Acad. Sci. USA* **1998**, *95*, 1336–1339. [CrossRef]

176. Hawkes, K. Genomic evidence for the evolution of human postmenopausal longevity. *Proc. Natl. Acad. Sci. USA* **2016**, *113*, 17–18. [CrossRef]

177. Ojeda-Granados, C.; Panduro, A.; Gonzalez-Aldaco, K.; Sepulveda-Villegas, M.; Rivera-Iñiguez, I.; Roman, S. Tailoring nutritional advice for Mexicans based on prevalence profiles of diet-related adaptive gene polymorphisms. *J. Pers. Med.* **2017**, *7*, 16. [CrossRef]

178. Singh, P.P.; Singh, M.; Mastana, S.S. APOE distribution in world populations with new data from India and the UK. *Ann. Hum. Biol.* **2006**, *33*, 279–308. [CrossRef]

179. Hu, P.; Qin, Y.H.; Jing, C.X.; Lu, L.; Hu, B.; Du, P.F. Does the geographical gradient of ApoE4 allele exist in China? A systemic comparison among multiple Chinese populations. *Mol. Biol. Rep.* **2011**, *38*, 489–494. [CrossRef]

180. Zekraoui, L.; Lagarde, J.P.; Raisonnier, A.; Gérard, N.; Aouizérate, A.; Lucotte, G. High frequency of the apolipoprotein E *4 allele in African pygmies and most of the African populations in sub-Saharan Africa. *Hum. Biol.* **1997**, *69*, 575–581.

181. Corbo, R.M.; Scacchi, R. Apolipoprotein E (*APOE*) allele distribution in the world. Is *APOE * 4* a 'thrifty' allele? *Ann. Hum. Genet.* **1999**, *63*, 301–310. [CrossRef] [PubMed]

182. Sazzini, M.; Gnecchi Ruscone, G.A.; Giuliani, C.; Sarno, S.; Quagliariello, A.; De Fanti, S.; Boattini, A.; Gentilini, D.; Fiorito, G.; Catanoso, M.; et al. Complex interplay between neutral and adaptive evolution shaped differential genomic background and disease susceptibility along the Italian peninsula. *Sci. Rep.* **2016**, *6*, 32513. [CrossRef] [PubMed]

183. Boattini, A.; Martinez-Cruz, B.; Sarno, S.; Harmant, C.; Useli, A.; Sanz, P.; Yang-Yao, D.; Manry, J.; Ciani, G.; Luiselli, D.; et al. Uniparental markers in Italy reveal a sex-biased genetic structure and different historical strata. *PLoS ONE* **2013**, *8*, e65441. [CrossRef]

184. Ye, K.; Gao, F.; Wang, D.; Bar-Yosef, O.; Keinan, A. Dietary adaptation of *FADS* genes in Europe varied across time and geography. *Nat. Ecol. Evol.* **2017**, *1*, 167. [CrossRef]

185. Buckley, M.T.; Racimo, F.; Allentoft, M.E.; Jensen, M.K.; Jonsson, A.; Huang, H.; Hormozdiari, F.; Sikora, M.; Marnetto, D.; Eskin, E.; et al. Selection in Europeans on fatty acid desaturases associated with dietary changes. *Mol. Biol. Evol.* **2017**, *34*, 1307–1318. [CrossRef]

186. Egert, S.; Rimbach, G.; Huebbe, P. ApoE genotype: From geographic distribution to function and responsiveness to dietary factors. *Proc. Nutr. Soc.* **2012**, *71*, 410–424. [CrossRef]

187. Graeser, A.-C.; Boesch-Saadatmandi, C.; Lippmann, J.; Wagner, A.E.; Huebbe, P.; Storm, N.; Höppner, W.; Wiswedel, I.; Gardemann, A.; Minihane, A.M.; et al. Nrf2-dependent gene expression is affected by the proatherogenic apoE4 genotype—studies in targeted gene replacement mice. *J. Mol. Med.* **2011**, *89*, 1027–1035. [CrossRef] [PubMed]

188. Huebbe, P.; Nebel, A.; Siegert, S.; Moehring, J.; Boesch-Saadatmandi, C.; Most, E.; Pallauf, J.; Egert, S.; Müller, M.J.; Schreiber, S.; et al. APOE ε4 is associated with higher vitamin D levels in targeted replacement mice and humans. *FASEB J.* **2011**, *25*, 3262–3270. [CrossRef]

189. Azevedo, O.G.R.; Bolick, D.T.; Roche, J.K.; Pinkerton, R.F.; Lima, A.A.M.; Vitek, M.P.; Warren, C.A.; Oriá, R.B.; Guerrant, R.L. Apolipoprotein E Plays a key role against cryptosporidial infection in transgenic undernourished mice. *PLoS ONE* **2014**, *9*. [CrossRef]

190. Huebbe, P.; Dose, J.; Schloesser, A.; Campbell, G.; Glüer, C.-C.; Gupta, Y.; Ibrahim, S.; Minihane, A.-M.; Baines, J.F.; Nebel, A.; et al. Apolipoprotein E (*APOE*) genotype regulates body weight and fatty acid utilization—Studies in gene-targeted replacement mice. *Mol. Nutr. Food Res.* **2015**, *59*, 334–343. [CrossRef]

191. Huebbe, P.; Lange, J.; Lietz, G.; Rimbach, G. Dietary β-carotene and lutein metabolism is modulated by the APOE genotype. *BioFactors* **2016**, *42*, 388–396. [CrossRef]

192. Luca, F.; Perry, G.H.; Di Rienzo, A. Evolutionary adaptations to dietary changes. *Annu. Rev. Nutr.* **2010**, *30*, 291–314. [CrossRef] [PubMed]

193. Finch, C.E.; Sapolsky, R.M. The evolution of Alzheimer disease, the reproductive schedule, and apoE isoforms. *Neurobiol. Aging* **1999**, *20*, 407–428. [CrossRef]

194. Schwarz, F.; Springer, S.A.; Altheide, T.K.; Varki, N.M.; Gagneux, P.; Varki, A. Human-specific derived alleles of *CD33* and other genes protect against postreproductive cognitive decline. *Proc. Natl. Acad. Sci.* **2016**, *113*, 74–79. [CrossRef] [PubMed]

195. van Exel, E.; Koopman, J.J.E.; van Bodegom, D.; Meij, J.J.; de Knijff, P.; Ziem, J.B.; Finch, C.E.; Westendorp, R.G.J. Effect of APOE ε4 allele on survival and fertility in an adverse environment. *PLoS ONE* **2017**, *12*, e0179497. [CrossRef]

196. Asgari, N.; Akbari, M.T.; Zare, S.; Babamohammadi, G. Positive association of apolipoprotein E4 polymorphism with recurrent pregnancy loss in Iranian patients. *J. Assist. Reprod. Genet.* **2013**, *30*, 265–268. [CrossRef] [PubMed]

197. Goodman, C.; Goodman, C.S.; Hur, J.; Jeyendran, R.S.; Coulam, C. The Association of apoprotien E polymorphisms with recurrent pregnancy loss. *Am. J. Reprod. Immunol.* **2009**, *61*, 34–38. [CrossRef]

198. Yenicesu, G.I.; Cetin, M.; Ozdemir, O.; Cetin, A.; Ozen, F.; Yenicesu, C.; Yildiz, C.; Kocak, N. A Prospective case–control study analyzes 12 thrombophilic gene mutations in Turkish couples with recurrent pregnancy loss. *Am. J. Reprod. Immunol.* **2010**, *63*, 126–136. [CrossRef] [PubMed]

199. Ozornek, H.; Ergin, E.; Jeyendran, R.S.; Ozay, A.T.; Pillai, D.; Coulam, C. Is apolipoprotien E codon 112 polymorphisms associated with recurrent pregnancy loss? *Am. J. Reprod. Immunol.* **2010**, *64*, 87–92. [CrossRef]

200. Ergin, E.; Jeyendran, R.S.; Özörnek, H.; Alev, Ö.; Pillai, M.D.; Coulam, C. Apolipoprotein E codon 112 polymorphisms is associated with recurrent pregnancy loss. *Fertil. Steril.* **2009**, *92*, S115. [CrossRef]

201. Vasunilashorn, S.; Finch, C.E.; Crimmins, E.M.; Vikman, S.A.; Stieglitz, J.; Gurven, M.; Kaplan, H.; Allayee, H. Inflammatory gene variants in the Tsimane, an indigenous Bolivian population with a high infectious load. *Biodemography Soc. Biol.* **2011**, *57*, 33–52. [CrossRef]

202. Trumble, B.C.; Stieglitz, J.; Blackwell, A.D.; Allayee, H.; Beheim, B.; Finch, C.E.; Gurven, M.; Kaplan, H. Apolipoprotein E4 is associated with improved cognitive function in Amazonian forager-horticulturalists with a high parasite burden. *FASEB J.* **2017**, *31*, 1508–1515. [CrossRef]

203. Tzourio, C.; Arima, H.; Harrap, S.; Anderson, C.; Godin, O.; Woodward, M.; Neal, B.; Bousser, M.-G.; Chalmers, J.; Cambien, F.; et al. APOE genotype, ethnicity, and the risk of cerebral hemorrhage. *Neurology* **2008**, *70*, 1322–1328. [CrossRef]

204. Zhang, M.; Gu, W.; Qiao, S.; Zhu, E.; Zhao, Q.; Lv, S. Apolipoprotein E gene polymorphism and risk for coronary heart disease in the Chinese population: A meta-analysis of 61 studies including 6634 cases and 6393 controls. *PLoS ONE* **2014**, *9*, e9546. [CrossRef]

205. Tang, M.-X.; Stern, Y.; Marder, K.; Bell, K.; Gurland, B.; Lantigua, R.; Andrews, H.; Feng, L.; Tycko, B.; Mayeux, R. The APOE-ε4 allele and the risk of Alzheimer disease among African Americans, Whites, and Hispanics. *JAMA* **1998**, *279*, 751–755. [CrossRef]

206. Wright, R.O.; Hu, H.; Silverman, E.K.; Tsaih, S.W.; Schwartz, J.; Bellinger, D.; Palazuelos, E.; Weiss, S.T.; Hernandez-Avila, M. Apolipoprotein E genotype predicts 24-Month Bayley scales infant development score. *Pediatr. Res.* **2003**, *54*, 819–825. [CrossRef]

207. Becher, J.-C.; Bell, J.E.; McIntosh, N.; Keeling, J.W. Distribution of apolipoprotein E alleles in a Scottish healthy newborn population. *Neonatology* **2005**, *88*, 164–167. [CrossRef]

208. Becher, J.; Keeling, J.W.; McIntosh, N.; Wyatt, B.; Bell, J. The distribution of apolipoprotein E alleles in Scottish perinatal deaths. *J. Med. Genet.* **2006**, *43*, 414–418. [CrossRef]

209. Becher, J.-C.; Keeling, J.W.; Bell, J.; Wyatt, B.; McIntosh, N. Apolipoprotein E e4 and its prevalence in early childhood death due to sudden infant death syndrome or to recognised causes. *Early Hum. Dev.* **2008**, *84*, 549–554. [CrossRef]

210. Farrer, L.A.; Cupples, L.A.; Haines, J.L.; Hyman, B.; Kukull, W.A.; Mayeux, R.; Myers, R.H.; Pericak-Vance, M.A.; Risch, N.; van Duijn, C.M. Effects of age, sex, and ethnicity on the association between apolipoprotein E genotype and Alzheimer disease: A meta-analysis. *JAMA* **1997**, *278*, 1349–1356. [CrossRef]

211. Bennet, A.M.; Angelantonio, E.D.; Ye, Z.; Wensley, F.; Dahlin, A.; Ahlbom, A.; Keavney, B.; Collins, R.; Wiman, B.; de Faire, U.; et al. Association of apolipoprotein E genotypes with lipid levels and coronary risk. *JAMA* **2007**, *298*, 1300–1311. [CrossRef]

212. Rougeron, V.; Woods, C.M.; Tiedje, K.E.; Bodeau-Livinec, F.; Migot-Nabias, F.; Deloron, P.; Luty, A.J.F.; Fowkes, F.J.I.; Day, K.P. Epistatic interactions between apolipoprotein E and Hemoglobin S genes in regulation of Malaria Parasitemia. *PLoS ONE* **2013**, *8*, e76924. [CrossRef] [PubMed]

213. Kulminski, A.M.; Raghavachari, N.; Arbeev, K.G.; Culminskaya, I.; Arbeeva, L.; Wu, D.; Ukraintseva, S.V.; Christensen, K.; Yashin, A.I. Protective role of the apolipoprotein E2 allele in age-related disease traits and survival: Evidence from the long life family study. *Biogerontology* **2016**, *17*, 893–905. [CrossRef]

214. Konishi, K.; Bhat, V.; Banner, H.; Poirier, J.; Joober, R.; Bohbot, V.D. APOE2 is associated with spatial navigational strategies and increased gray matter in the hippocampus. *Front. Hum. Neurosci.* **2016**, *10*. [CrossRef]

215. Cashdan, E. Biogeography of human infectious diseases: A global historical analysis. *PLoS ONE* **2014**, *9*. [CrossRef]

216. Guernier, V.; Hochberg, M.E.; Guégan, J.-F. Ecology drives the worldwide distribution of human diseases. *PLoS Biol.* **2004**, *2*, e141. [CrossRef]

217. Barger, S.W.; Harmon, A.D. Microglial activation by Alzheimer amyloid precursor protein and modulation by apolipoprotein E. *Nature* **1997**, *388*, 878–881. [CrossRef] [PubMed]

218. Bonacina, F.; Coe, D.; Wang, G.; Longhi, M.P.; Baragetti, A.; Moregola, A.; Garlaschelli, K.; Uboldi, P.; Pellegatta, F.; Grigore, L.; et al. Myeloid apolipoprotein E controls dendritic cell antigen presentation and T cell activation. *Nat. Commun.* **2018**, *9*, 3083. [CrossRef]

219. Kelly, M.E.; Clay, M.A.; Mistry, M.J.; Hsieh-Li, H.-M.; Harmony, J.A.K. Apolipoprotein E inhibition of proliferation of mitogen-activated T lymphocytes: Production of interleukin 2 with reduced biological activity. *Cell. Immunol.* **1994**, *159*, 124–139. [CrossRef]

220. Picardi, A.; Gentilucci, U.V.; Bambacioni, F.; Galati, G.; Spataro, S.; Mazzarelli, C.; D'Avola, D.; Fiori, E.; Riva, E. Lower schooling, higher hepatitis C virus prevalence in Italy: An association dependent on age. *J. Clin. Virol.* **2007**, *40*, 168–170. [CrossRef]

221. Corbo, R.M.; Scacchi, R.; Mureddu, L.; Mulas, G.; Alfano, G. Apolipoprotein E polymorphism in Italy investigated in native plasma by a simple polyacrylamide gel isoelectric focusing technique. Comparison with frequency data of other European populations. *Ann. Hum. Genet.* **1995**, *59*, 197–209. [CrossRef]

222. Kuhlmann, I.; Minihane, A.M.; Huebbe, P.; Nebel, A.; Rimbach, G. Apolipoprotein E genotype and hepatitis C, HIV and herpes simplex disease risk: A literature review. *Lipids Health Dis.* **2010**, *9*, 8. [CrossRef]

223. Soares, H.D.; Potter, W.Z.; Pickering, E.; Kuhn, M.; Immermann, F.W.; Shera, D.M.; Ferm, M.; Dean, R.A.; Simon, A.J.; Swenson, F.; et al. Biomarkers associated with the apolipoprotein E genotype and Alzheimer disease. *Arch. Neurol.* **2012**, *69*, 1310–1317. [CrossRef]

224. Gale, S.C.; Gao, L.; Mikacenic, C.; Coyle, S.M.; Rafaels, N.; Murray, T.; Madenspacher, J.H.; Draper, D.W.; Ge, W.; Aloor, J.J.; et al. APOε 4 is associated with enhanced in vivo innate immune responses in humans. *J. Allergy Clin. Immunol.* **2014**, *134*, 127–134. [CrossRef]

225. Oosten, M.V.; Rensen, P.C.N.; Amersfoort, E.S.V.; Eck, M.V.; Dam, A.-M.V.; Brevé, J.J.P.; Vogel, T.; Panet, A.; Berkel, T.J.C.V.; Kuiper, J. Apolipoprotein E protects against bacterial lipopolysaccharide-induced lethality a new therapeutic approach to treat gram-negative sepsis. *J. Biol. Chem.* **2001**, *276*, 8820–8824. [CrossRef]

226. Liu, B.; Zhang, Y.; Wang, R.; An, Y.; Gao, W.; Bai, L.; Li, Y.; Zhao, S.; Fan, J.; Liu, E. Western diet feeding influences gut microbiota profiles in apoE knockout mice. *Lipids Health Dis.* **2018**, *17*, 159. [CrossRef]

227. Kasahara, K.; Tanoue, T.; Yamashita, T.; Yodoi, K.; Matsumoto, T.; Emoto, T.; Mizoguchi, T.; Hayashi, T.; Kitano, N.; Sasaki, N.; et al. Commensal bacteria at the crossroad between cholesterol homeostasis and chronic inflammation in atherosclerosis. *J. Lipid Res.* **2017**, *58*, 519–528. [CrossRef]

228. Saita, D.; Ferrarese, R.; Foglieni, C.; Esposito, A.; Canu, T.; Perani, L.; Ceresola, E.R.; Visconti, L.; Burioni, R.; Clementi, M.; et al. Adaptive immunity against gut microbiota enhances apoE-mediated immune regulation and reduces atherosclerosis and western-diet-related inflammation. *Sci. Rep.* **2016**, *6*, 29353. [CrossRef]

229. Schneeberger, M.; Everard, A.; Gómez-Valadés, A.G.; Matamoros, S.; Ramírez, S.; Delzenne, N.M.; Gomis, R.; Claret, M.; Cani, P.D. *Akkermansia muciniphila* inversely correlates with the onset of inflammation, altered adipose tissue metabolism and metabolic disorders during obesity in mice. *Sci. Rep.* **2015**, *5*, 16643. [CrossRef]

230. Li, J.; Lin, S.; Vanhoutte, P.M.; Woo, C.W.; Xu, A. *Akkermansia Muciniphila* protects against atherosclerosis by preventing metabolic endotoxemia-induced inflammation in ApoE−/− mice. *Circulation* **2016**, *133*, 2434–2446. [CrossRef]

231. Zhao, S.; Liu, W.; Wang, J.; Shi, J.; Sun, Y.; Wang, W.; Ning, G.; Liu, R.; Hong, J. *Akkermansia muciniphila* improves metabolic profiles by reducing inflammation in chow diet-fed mice. *J. Mol. Endocrinol.* **2017**, *58*, 1–14. [CrossRef] [PubMed]

232. Yashin, A.I.; Wu, D.; Arbeev, K.G.; Ukraintseva, S.V. Polygenic effects of common single-nucleotide polymorphisms on life span: When association meets causality. *Rejuvenation Res.* **2012**, *15*, 381–394. [CrossRef] [PubMed]

233. Boyle, E.A.; Li, Y.I.; Pritchard, J.K. An expanded view of complex traits: From polygenic to omnigenic. *Cell* **2017**, *169*, 1177–1186. [CrossRef] [PubMed]

234. Gnecchi-Ruscone, G.A.; Abondio, P.; De Fanti, S.; Sarno, S.; Sherpa, M.G.; Sherpa, P.T.; Marinelli, G.; Natali, L.; Di Marcello, M.; Peluzzi, D.; et al. Evidence of polygenic adaptation to high altitude from Tibetan and Sherpa genomes. *Genome Biol. Evol.* **2018**, *10*, 2919–2930. [CrossRef]

235. McClellan, J.; King, M.-C. Genetic heterogeneity in human disease. *Cell* **2010**, *141*, 210–217. [CrossRef] [PubMed]

236. Wang, B.; Bao, S.; Zhang, Z.; Zhou, X.; Wang, J.; Fan, Y.; Zhang, Y.; Li, Y.; Chen, L.; Jia, Y.; et al. A rare variant in *MLKL* confers susceptibility to ApoE ε4-negative Alzheimer's disease in Hong Kong Chinese population. *Neurobiol. Aging* **2018**, *68*, 160.e1–160.e7. [CrossRef] [PubMed]

237. *Evolutionary Thinking in Medicine: From Research to Policy and Practice*; Alvergne, A.; Jenkinson, C.; Faurie, C. (Eds.) Springer International Publishing: Basel, Switzerland, 2016.

238. Allentoft, M.E.; Sikora, M.; Sjögren, K.-G.; Rasmussen, S.; Rasmussen, M.; Stenderup, J.; Damgaard, P.B.; Schroeder, H.; Ahlström, T.; Vinner, L.; et al. Population genomics of Bronze Age Eurasia. *Nature* **2015**, *522*, 167–172. [CrossRef]

239. Olalde, I.; Brace, S.; Allentoft, M.E.; Armit, I.; Kristiansen, K.; Booth, T.; Rohland, N.; Mallick, S.; Szécsényi-Nagy, A.; Mittnik, A.; et al. The Beaker phenomenon and the genomic transformation of northwest Europe. *Nature* **2018**, *555*, 190–196. [CrossRef] [PubMed]

240. Olalde, I.; Schroeder, H.; Sandoval-Velasco, M.; Vinner, L.; Lobón, I.; Ramirez, O.; Civit, S.; García Borja, P.; Salazar-García, D.C.; Talamo, S.; et al. A common genetic origin for early farmers from Mediterranean Cardial and Central European LBK cultures. *Mol. Biol. Evol.* **2015**, *32*, 3132–3142. [CrossRef]

241. Olalde, I.; Allentoft, M.E.; Sánchez-Quinto, F.; Santpere, G.; Chiang, C.W.K.; DeGiorgio, M.; Prado-Martinez, J.; Rodríguez, J.A.; Rasmussen, S.; Quilez, J.; et al. Derived immune and ancestral pigmentation alleles in a 7000-year-old Mesolithic European. *Nature* **2014**, *507*, 225–228. [CrossRef]

242. Mathieson, I.; Alpaslan-Roodenberg, S.; Posth, C.; Szécsényi-Nagy, A.; Rohland, N.; Mallick, S.; Olalde, I.; Broomandkhoshbacht, N.; Candilio, F.; Cheronet, O.; et al. The genomic history of southeastern Europe. *Nature* **2018**, *555*, 197–203. [CrossRef] [PubMed]

243. Mathieson, I.; Lazaridis, I.; Rohland, N.; Mallick, S.; Patterson, N.; Roodenberg, S.A.; Harney, E.; Stewardson, K.; Fernandes, D.; Novak, M.; et al. Genome-wide patterns of selection in 230 ancient Eurasians. *Nature* **2015**, *528*, 499–503. [CrossRef] [PubMed]

244. Jones, E.R.; Gonzalez-Fortes, G.; Connell, S.; Siska, V.; Eriksson, A.; Martiniano, R.; McLaughlin, R.L.; Gallego Llorente, M.; Cassidy, L.M.; Gamba, C.; et al. Upper Palaeolithic genomes reveal deep roots of modern Eurasians. *Nat. Commun.* **2015**, *6*, 8912. [CrossRef]

245. Haak, W.; Lazaridis, I.; Patterson, N.; Rohland, N.; Mallick, S.; Llamas, B.; Brandt, G.; Nordenfelt, S.; Harney, E.; Stewardson, K.; et al. Massive migration from the steppe was a source for Indo-European languages in Europe. *Nature* **2015**, *522*, 207–211. [CrossRef] [PubMed]

246. Lipson, M.; Szécsényi-Nagy, A.; Mallick, S.; Pósa, A.; Stégmár, B.; Keerl, V.; Rohland, N.; Stewardson, K.; Ferry, M.; Michel, M.; et al. Parallel palaeogenomic transects reveal complex genetic history of early European farmers. *Nature* **2017**, *551*, 368–372. [CrossRef] [PubMed]

247. Lazaridis, I.; Patterson, N.; Mittnik, A.; Renaud, G.; Mallick, S.; Kirsanow, K.; Sudmant, P.H.; Schraiber, J.G.; Castellano, S.; Lipson, M.; et al. Ancient human genomes suggest three ancestral populations for present-day Europeans. *Nature* **2014**, *513*, 409–413. [CrossRef] [PubMed]

248. Lazaridis, I.; Mittnik, A.; Patterson, N.; Mallick, S.; Rohland, N.; Pfrengle, S.; Furtwängler, A.; Peltzer, A.; Posth, C.; Vasilakis, A.; et al. Genetic origins of the Minoans and Mycenaeans. *Nature* **2017**, *548*, 214–218. [CrossRef] [PubMed]

249. Schiffels, S.; Haak, W.; Paajanen, P.; Llamas, B.; Popescu, E.; Loe, L.; Clarke, R.; Lyons, A.; Mortimer, R.; Sayer, D.; et al. Iron Age and Anglo-Saxon genomes from East England reveal British migration history. *Nat. Commun.* **2016**, *7*, 10408. [CrossRef]

250. Martiniano, R.; Caffell, A.; Holst, M.; Hunter-Mann, K.; Montgomery, J.; Müldner, G.; McLaughlin, R.L.; Teasdale, M.D.; van Rheenen, W.; Veldink, J.H.; et al. Genomic signals of migration and continuity in Britain before the Anglo-Saxons. *Nat. Commun.* **2016**, *7*, 10326. [CrossRef]

251. Broushaki, F.; Thomas, M.G.; Link, V.; López, S.; van Dorp, L.; Kirsanow, K.; Hofmanová, Z.; Diekmann, Y.; Cassidy, L.M.; Díez-del-Molino, D.; et al. Early Neolithic genomes from the eastern Fertile Crescent. *Science* **2016**, *353*, 499–503. [CrossRef] [PubMed]

252. Cassidy, L.M.; Martiniano, R.; Murphy, E.M.; Teasdale, M.D.; Mallory, J.; Hartwell, B.; Bradley, D.G. Neolithic and Bronze Age migration to Ireland and establishment of the insular Atlantic genome. *Proc. Natl. Acad. Sci. USA* **2016**, *113*, 368–373. [CrossRef] [PubMed]

253. Raghavan, M.; Skoglund, P.; Graf, K.E.; Metspalu, M.; Albrechtsen, A.; Moltke, I.; Rasmussen, S.; Stafford, T.W., Jr.; Orlando, L.; Metspalu, E.; et al. Upper Palaeolithic Siberian genome reveals dual ancestry of Native Americans. *Nature* **2014**, *505*, 87–91. [CrossRef] [PubMed]

254. Fu, Q.; Li, H.; Moorjani, P.; Jay, F.; Slepchenko, S.M.; Bondarev, A.A.; Johnson, P.L.F.; Aximu-Petri, A.; Prüfer, K.; de Filippo, C.; et al. Genome sequence of a 45,000-year-old modern human from western Siberia. *Nature* **2014**, *514*, 445–449. [CrossRef] [PubMed]

255. Günther, T.; Valdiosera, C.; Malmström, H.; Ureña, I.; Rodriguez-Varela, R.; Sverrisdóttir, Ó.O.; Daskalaki, E.A.; Skoglund, P.; Naidoo, T.; Svensson, E.M.; et al. Ancient genomes link early farmers from Atapuerca in Spain to modern-day Basques. *Proc. Natl. Acad. Sci. USA* **2015**, *112*, 11917–11922. [CrossRef] [PubMed]

256. Hofmanová, Z.; Kreutzer, S.; Hellenthal, G.; Sell, C.; Diekmann, Y.; Díez-del-Molino, D.; van Dorp, L.; López, S.; Kousathanas, A.; Link, V.; et al. Early farmers from across Europe directly descended from Neolithic Aegeans. *Proc. Natl. Acad. Sci. USA* **2016**, *113*, 6886–6891. [CrossRef] [PubMed]

257. Kılınç, G.M.; Omrak, A.; Özer, F.; Günther, T.; Büyükkarakaya, A.M.; Bıçakçı, E.; Baird, D.; Dönertaş, H.M.; Ghalichi, A.; Yaka, R.; et al. The Demographic development of the First Farmers in Anatolia. *Curr. Biol.* **2016**, *26*, 2659–2666. [CrossRef] [PubMed]

258. Keller, A.; Graefen, A.; Ball, M.; Matzas, M.; Boisguerin, V.; Maixner, F.; Leidinger, P.; Backes, C.; Khairat, R.; Forster, M.; et al. New insights into the Tyrolean Iceman's origin and phenotype as inferred by whole-genome sequencing. *Nat. Commun.* **2012**, *3*, 698. [CrossRef] [PubMed]

259. Nesse, R.M. Evolution: Medicine's most basic science. *Lancet* **2008**, *372*, S21–S27. [CrossRef]

260. Nesse, R.M. Ten questions for evolutionary studies of disease vulnerability. *Evol. Appl.* **2011**, *4*, 264–277. [CrossRef] [PubMed]

261. Nesse, R.M.; Bergstrom, C.T.; Ellison, P.T.; Flier, J.S.; Gluckman, P.; Govindaraju, D.R.; Niethammer, D.; Omenn, G.S.; Perlman, R.L.; Schwartz, M.D.; et al. Making evolutionary biology a basic science for medicine. *Proc. Natl. Acad. Sci. USA* **2010**, *107*, 1800–1807. [CrossRef] [PubMed]

262. Wells, J.C.K.; Nesse, R.M.; Sear, R.; Johnstone, R.A.; Stearns, S.C. Evolutionary public health: Introducing the concept. *Lancet* **2017**, *390*, 500–509. [CrossRef]

263. Randolph, M.; Nesse, M.D.; George, C.W. Why We Get Sick. Available online: https://www.penguinrandomhouse.com/books/120768/why-we-get-sick-by-randolph-m-nesse/9780679746744 (accessed on 1 March 2019).

GCAT
TACG
GCAT
genes

MDPI

Article

Lifespans of Twins: Does Zygosity Matter?

Jacob Hjelmborg [1,2,*], Pia Larsen [1], Jaakko Kaprio [3,4], Matt McGue [1,5], Thomas Scheike [6], Philip Hougaard [1,7] and Kaare Christensen [1,2]

1   Department of Epidemiology and Biostatistics, University of Southern Denmark, DK-5000 Odense, Denmark; plarsen@health.sdu.dk (P.L.); mcgue001@umn.edu (M.M.); phou@lundbeck.com (P.H.); kchristensen@health.sdu.dk (K.C.)
2   The Danish Twin Registry, University of Southern Denmark, DK-5000 Odense, Denmark
3   Department of Public Health, University of Helsinki, FI-00014 Helsinki, Finland; jaakko.kaprio@helsinki.fi
4   Institute for Molecular Medicine FIMM, University of Helsinki, FI-00014 Helsinki, Finland
5   Department of Psychology, University of Minnesota, Minneapolis, MN 55455, USA
6   Department of Biostatistics, University of Copenhagen, DK-1014 Copenhagen K, Denmark; bhd252@ku.dk
7   Biometric Division, Lundbeck, DK-2500 Valby, Denmark
*   Correspondence: jhjelmborg@health.sdu.dk; Tel.: +45-6550-3075

Received: 15 January 2019; Accepted: 18 February 2019; Published: 20 February 2019

Abstract: Studies with twins provide fundamental insights to lifespans of humans. We aim to clarify if monozygotic and dizygotic twin individuals differ in lifespan, that is, if zygosity matters. We investigate whether a possible difference in mortality after infancy between zygosities is stable in different age cohorts, and whether the difference remains when twins with unknown zygosity are taken into account. Further, we compare the distribution of long-livers, that is, the upper-tail of the lifespan distribution, between monozygotic and same-sex dizygotic twin individuals. The Danish Twin Registry provides a nationwide cohort of 109,303 twins born during 1870 to 1990 with valid vital status. Standard survival analysis is used to compare mortality in monozygotic and dizygotic twin individuals and twin individuals with unknown zygosity. The mortality of monozygotic and dizygotic twin individuals differs slightly after taking into consideration effects of birth- and age-cohorts, gender differences, and that twins are paired. However, no substantial nor systematic differences remain when taking twins with unknown zygosity into account. Further, the distribution of long-livers is very similar by zygosity, suggesting the same mortality process. The population-based and oldest twin cohort ever studied suggests that monozygotic and dizygotic twins have similar lifespans.

Keywords: lifespan; mortality; twins; zygosity; unknown zygosity; cumulative incidence curves; age-stratification; long-livers

1. Introduction

One of the advantages of using twins in quantitative genetic modelling is the assumption of comparability (of trait means and distribution of variances) of the monozygotic (MZ) and dizygotic (DZ) twins with respect to the studied trait or disease when considering the twins as individuals [1]. However, for studies on disease onset and other age-dependent traits, an important factor to study is the relative survival of MZ twins compared to DZ twins [2]. One may speculate whether MZ twins in general show better survival in comparison to same-sex and opposite-sex DZ twins as an effect of greater closeness and social support. On the other hand, MZ twins show lower birth weights, higher risk of prematurity and greater neonatal mortality than DZ twins, which may affect long term survival [3]. Genetic and environmental effects on mortality are well studied [4,5], showing a modestly higher concordance of overall mortality in MZ twin pairs than DZ twin pairs. However, the effect of zygosity on concordance in mortality might integrate social factors beneficial for a longer life. It may be difficult to disentangle such factors, and many studies points towards little or no difference in twins as individuals across zygosity for many outcomes related to survival [6,7].

Compared to the background population, twin individuals have higher perinatal and infant mortality [8], however the relatively high mortality in twins occurs soon after birth [2], and after the age of six mortality of twins is similar to that of the background population [6]. A recent study on Danish twin birth cohorts from 1870–1900, left truncated at age 10, suggested that MZ twin individuals have better survival than same-sex DZ twin individuals at nearly all ages [9]. The paper applied a two-process model, assuming independence between all subjects, and partitioned mortality into an intrinsic process, where death follows from cumulative and incremental degradation of survival capacity, and an extrinsic process, where death results from an acute environmental challenge [9]. No formal statistical tests were made, but explorative results indicated that MZ female twins may have better overall survival than same-sex DZ female twins, due to greater extrinsic survival. Monozygotic male twins may have better overall survival than same-sex DZ male twins, where the improved survival was ascribed to greater extrinsic survival until age 65–70, and to greater intrinsic survival from age 65–70. Based on these results, we seek here to study rigorously mortality after infancy in MZ, same-sex DZ, and opposite-sex DZ twin individuals, accounting for the dependency within twin pairs, in birth cohorts ranging from 1870 until 1990 and using the most recent follow-up data on vital status in the Danish Twin Registry [10,11].

The Danish Twin Registry was established in 1954, identifying twin pairs prospectively as well as retrospectively from church records and classifying zygosity through questionnaires and interviews with surviving twins and family [11,12]. The register includes Danish twin pairs from birth cohorts 1870 onwards and is regarded almost complete from birth cohort 1960 [11,13]. For some same-sex twin pairs the zygosity is not determined, in particular in the early birth cohorts and among twin pairs with early mortality. In the early birth cohorts from 1870 to 1960, more same sex twins with early mortality are likely to be registered as unknown zygosity (UZ) due to the ascertainment method, and consequently, mortality of UZ twins from birth cohorts 1870–1960 may be expected to be higher than for twins registered as MZ or DZ. From birth cohorts 1960, it may be assumed that the registration of UZ twins due to death after infancy is negligible [10]. Although the true zygosity of UZ twins is unknown, it is expected that UZ twin pairs are a mixture of MZ and same-sex DZ twin pairs.

The objectives of this study are to compare mortality after infancy in MZ, same-sex DZ, and opposite-sex DZ twin individuals, to examine whether possible differences in survival between zygosities are stable in different and age cohorts, and to investigate differences in mortality in MZ, same-sex DZ, and opposite-sex DZ under different assumptions on the true zygosity of UZ twin pairs. A further objective was to explore the upper-end of the lifespan distribution in MZ and same-sex DZ twin individuals.

2. Materials and Methods

2.1. The Danish Twin Registry

As the first nationwide twin registry, the Danish Twin Registry was established in 1954, identifying twin pairs prospectively as well as retrospectively from church records of all 2200 Danish parishes and all calendar years 1870–1930. Also, regional population registers and other public sources were used to identify twins and close relatives of twins. All twins, or their closest relatives, were sent questionnaires, including questions on similarity between the twins to determine the zygosity of same-sex twin pairs [6,11,12,14,15]. If one or both twins had died or emigrated before the age of six, the twin pairs were not followed up. Twins, who did not reply to the questionnaires, as well as a minority providing inconsistent responses, were classified as UZ [10]. Later comparisons of the zygosities of same-sex twins determined from the questionnaires with results determined from blood samples indicated that more than 95% of twin pairs were classified correctly [16].

For the birth cohorts 1930–1990, the ascertainment procedure was based on the civil registration system introduced in 1968 and hence the criterion for inclusion was survival of both twins until 2 April 1968, which resulted in complete registration of all twins born from that date [10].

2.2. Study Population

The study includes population-based birth cohorts from the Danish Twin Registry including information about mortality. The Danish Twin Register comprises total of 126,489 individuals from multiple births from birth cohorts 1870–1990. Excluding triplets, quadruplets, twin pairs with inconsistent zygosities, twin individuals with invalid or missing vital status, and twin individuals who died before age six, leaves a total of 109,303 twin individuals from 57,313 twin pairs. For our primary analyses on differences in mortality between zygosities, we use all twin individuals with known zygosity from birth cohorts 1870–1990 (*n* = 96,338), and for our secondary analyses we include UZ twins and restrict the birth cohorts to 1961–1990 (*n* = 39,504) where there is virtually complete vital status follow-up of the twin individuals [10] (Figure 1).

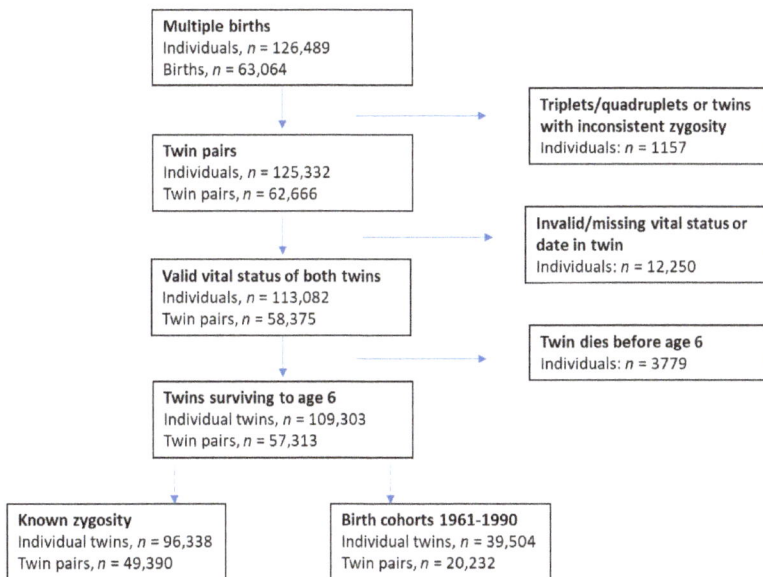

Figure 1. Flowchart.

The project was approved, and data were provided by, the Danish Twin Registry (ref. no. 17/64840) in an anonymized fashion. Access to the data requires application to this registry. According to Danish law, no ethical approval is required for registry-based studies.

2.3. Statistical Analyses

Individual twins were excluded in case of missing data on vital status or date of vital status, or in case of death before age six years old (Figure 1). For the general population, it is well known that life expectancy has improved over the period 1870–1990. To be able to handle this period-effect and the differences in ascertainment methods, the birth cohorts are divided into four cohorts within which the same ascertainment method was used, and within which mortality rates are assumed to be homogeneous in the background population: In the first two cohorts, church records were used for identification of twins; this cohort was further divided into two separate cohorts, due to the pronounced increase in life expectancy in the Danish background population during this period, increasing from an average life expectancy in males between 46–49 years in 1870–1900 to an average life expectancy in males between 53–61 years in 1901–1930 (StatBank Denmark, Statistics Denmark). In the third cohort, including birth cohorts 1931–1960, the ascertainment of twins was based on survival of both twins until April 1968 and the average life expectancy in the background male population ranged between 62–70 years. From 1960, virtually all twins surviving infancy were included, and the average life expectancy in the background male population was between 70–72. Descriptive analyses were conducted for each cohort separately. Mortality was compared between zygosities (MZ: monozygotic, SSDZ: same-sex dizygotic, OSDZ: opposite-sex dizygotic) separately within each cohort using Kaplan-Meier plots and cumulative incidence curves [17] of death by age, stratified on sex. Due to non-proportional hazards and statistically significant interaction between zygosity and age, the models were stratified on three age-intervals (0–50 years, 50–75 years and 75+ years) by introducing a time-varying interaction term with cut-points at ages 50 years and 75 years. Within each age-interval, the proportional hazards assumptions were satisfied when assessed using Schoenfeld residual tests. To combine survival analyses across the four cohorts, the baseline hazards were stratified on the cohorts in the age-stratified Cox proportional hazard models. In analyses on both sexes combined, the baseline hazards were further stratified on sex. The mortality analyses were conducted on twins regarded as individuals. However, as the two twins within a twin pair are not independent, cluster robust standard errors were used in the statistical models to account for the lack of independence between observations [18]. All analyses were conducted on both sexes combined and for each sex separately.

Registration of twins surviving infancy is assumed to be complete in the Danish Twin register from 1960 onwards. Therefore, the fourth cohort (birth cohorts 1961–1990) was used in three sensitivity analyses including all twin individuals with known or unknown zygosities. The cohort comprises a total of 39,504 twin individuals of which 5178 have UZ. In the first sensitivity analysis, all UZ twin pairs were treated as MZ, in the second UZ twin pairs were randomized 1:1 as either MZ or SSDZ twins, and in the third, UZ twin pairs were randomized 1:2 as either MZ or SSDZ twins. In the cohort 1961–1990, the follow-up time was at most 56 years (from 1 January 1960 until end of follow-up 1 October 2016), and most twin individuals were alive at the time of follow-up. Age-stratified (0–50 years and 50+ years) Cox proportional hazard regression analyses with cluster robust standard errors were conducted. In analyses on both sexes combined, the baseline hazards were stratified on sex. The analyses were conducted on both sexes combined and for each sex separately. Assumptions on proportional hazards were assessed using Schoenfeld residual tests and were satisfied for both sexes and in both age-groups.

Finally, we explored whether the distribution of the lifespans of long-livers, that is, twin individuals reaching very high ages, differ between MZ and SSDZ twin individuals, by comparing the upper-tail of the lifespan distributions. To do this, the generalized extreme value distribution (GEV) was applied to model the upper-tail of the lifespan distribution for twins born between 1870–1930 [19]. The GEV is a three-parameter model with a location parameter μ, estimating the center of the lifespan distribution for twin individuals, a scale parameter σ, estimating the deviations around the location parameter, and finally a shape parameter ξ, estimating the heaviness of the upper-tail lifespan, i.e., the distribution of lifespan among the long-livers: twins reaching very high ages. An increasing shape parameter corresponds to a higher probability of reaching very long lifespans. A negative value of the shape parameter indicates a light upper-tail, i.e., that there is a final limit to the highest possible lifespan, which is in accordance with human lifespans [20]. The shape parameter of the GEV to be reported for MZ and SSDZ for each gender was estimated by the moments method and inference was obtained by parametric bootstrap using the R-package ExtRemes [21].

3. Results

Characteristics of twin individuals with known zygosities in each of the four cohorts are shown in Table 1. The age at follow-up (FU) was lower in the third and fourth cohorts due to censoring at end of the study period (1 October 2016). In the third cohort, the proportions of MZ twins and females were lower than in the other cohorts. In the second cohort, the proportion of OSDZ twins was lower than in the other cohorts.

Table 1. Population characteristics of individual twins, surviving to age 6, in the four cohorts, excluding UZ.

	All cohorts 1870–1990	Cohort 1 1870–1900	Cohort 2 1901–1930	Cohort 3 1931–1960	Cohort 4 1961–1990
Total, *n* (pairs)	96,338 (49,390)	9037 (4907)	15,645 (8060)	37,330 (19,114)	34,326 (17,309)
Age at FU [a], mean (SD)	58.5 (18.3)	69.0 (19.8)	74.0 (16.8)	65.3 (11.2)	41.2 (9.2)
Sex, *n* (%)					
Male	49,100 (51.0)	4496 (49.8)	7480 (47.8)	20,133 (53.9)	16,991 (49.5)
Female	47,238 (49.0)	4541 (50.3)	8165 (52.2)	17,197 (46.1)	17,335 (50.5)
Zygosity, *n* (%)					
MZ	23,888 (24.8)	2247 (24.9)	4613 (29.5)	7122 (19.1)	9906 (28.9)
SSDZ	39,728 (41.2)	4029 (44.6)	9156 (58.5)	14,440 (38.9)	12,103 (35.3)
OSDZ	32,722 (34.0)	2761 (30.6)	1876 (12.0)	15,768 (42.2)	12,317 (35.9)
Dead during FU [a], *n* (%)					
No/censored	62,459 (64.8)	208 (2.3)	959 (6.1)	27,806 (74.5)	33,486 (97.6)
Yes	33,879 (35.2)	8829 (97.7)	14,686 (93.9)	9524 (25.5)	840 (2.4)

[a] FU is defined as time until death, censoring (emigration) or end of study period (1 October 2016). SD: standard deviation; MZ: monozygotic; SSDZ: same-sex dizygotic; OSDZ: opposite-sex dizygotic.

Kaplan-Meier survival curves indicate non-proportional hazards in the two first cohorts for both male and female twins (Supplementary Figures S1 and S2). For younger ages, the mortality of OSDZ twin individuals appears much higher than both MZ and SSDZ twin individuals in the first two cohorts, while at higher ages the survival is similar for all three zygosities.

In all three age-intervals, OSDZ twin individuals have significantly higher mortality than MZ twin individuals for both males and females (Table 2). For males, SSDZ twin individuals have significantly higher mortality than MZ twin individuals in all three age-intervals, and for females SSDZ twin individuals have significantly higher mortality than MZ twin individuals in the two younger age intervals. For both males and females, the difference in mortality between MZ and DZ twin individuals is largest in the youngest age interval (0–50 years) and weakens with age (Table 2).

Secondary analyses were conducted on the birth cohorts 1961–1990 including UZ twin pairs. Population characteristics of the twin individuals are shown in Table 3.

Kaplan-Meier survival curves (Figure S3) suggest a higher mortality in UZ twin individuals than twin individuals with known zygosity, mainly among male twins. Age stratified analyses on the birth cohorts 1961–1990 in Table 4 show that the increased mortality among UZ male twin individuals mainly exists in the younger age group (up to 50 years of age), while there are no statistically significant differences in mortality in twin individuals aged above 50.

Table 2. Associations between mortality and zygosity in individual twins from all birth cohorts 1870–1990 and surviving to age 6; among individual twins at ages up to 50 years, between 50–75 years, and from 75 years.

	All Ages, $n = 96,338$ [a]	Ages $\leq$50, $n = 96,338$ [a]	Ages 51–75, $n = 65,107$ [a]	Ages >75, $n = 20,839$ [a]
Both sexes	HR (95%-CI)	HR (95%-CI)	HR (95%-CI)	HR (95%-CI)
Zygosity, *n* (%)				
MZ	Ref.	Ref.	Ref.	Ref.
SSDZ	1.09 (1.06, 1.12) ***	1.19 (1.11, 1.29) ***	1.11 (1.06, 1.17) ***	1.04 (1.00, 1.09) *
OSDZ	1.17 (1.13, 1.21) ***	1.53 (1.41, 1.66) ***	1.12 (1.06, 1.18) ***	1.09 (1.04, 1.15) ***
Males				
Zygosity, *n* (%)				
MZ	Ref.	Ref.	Ref.	Ref.
SSDZ	1.11 (1.07, 1.16) ***	1.23 (1.11, 1.36) ***	1.11 (1.04, 1.18) ***	1.07 (1.00, 1.13) *
OSDZ	1.16 (1.11, 1.21) ***	1.44 (1.30, 1.60) ***	1.10 (1.03, 1.18) **	1.11 (1.03, 1.19) **
Females				
Zygosity, *n* (%)				
MZ	Ref.	Ref.	Ref.	Ref.
SSDZ	1.07 (1.03, 1.12) **	1.20 (1.11, 1.30) ***	1.12 (1.07, 1.17) ***	1.04 (1.00, 1.08)
OSDZ	1.18 (1.13, 1.24) ***	1.52 (1.40, 1.64) ***	1.11 (1.06, 1.17) ***	1.09 (1.04, 1.15) ***

[a] Baseline hazard stratified on cohort group (analyses on both sexes stratified on sex as well). HR: hazard ratio; CI: confidence interval; * $p < 0.05$, ** $p < 0.01$, *** $p < 0.001$.

Table 3. Population characteristics of individual twins, surviving to age 6, birth cohorts 1961–1990.

	All Subjects 1961–1990	Males 1961–1990	Females 1961–1990
Total, *n* (pairs)	39,504 (20,232)	19,964 (13,266)	19,540 (13,000)
Age at FU [a], mean (SD)	40.3 (9.6)	40.2 (9.5)	40.5 (9.6)
Sex, *n* (%)			
Male	19,964 (50.5)	19,964 (100.0)	-
Female	19,540 (49.5)	-	19,540 (100.0)
Zygosity, *n* (%)			
MZ	9906 (25.1)	4689 (23.5)	5217 (26.7)
SSDZ	12,103 (30.6)	6159 (30.9)	5944 (30.4)
OSDZ	12,317 (31.2)	6143 (30.8)	6174 (31.6)
UZ	5178 (13.1)	2973 (14.9)	2205 (11.3)
Dead during FU [a], *n* (%)			
No	38,519 (97.5)	19,334 (96.8)	19,185 (98.2)
Yes	985 (2.5)	630 (3.2)	355 (1.8)

[a] FU is defined as time until death, censoring (emigration) or end of study period (1 October 2016).

Table 4. Associations between mortality and zygosity in individual twins, birth cohorts 1961–1990, surviving to age 6; among individual twins at ages up to 50 years, and from 50 years.

	All Ages, n = 39,504	Ages ≤50, n = 39,504	Ages >50, n = 8711
Both sexes [a]	HR (95%-CI)	HR (95%-CI)	HR (95%-CI)
Zygosity			
MZ	Ref.	Ref.	Ref.
SSDZ	1.07 (0.89, 1.13)	1.10 (0.91, 1.34)	0.80 (0.46, 1.40)
OSDZ	1.21 (1.01, 1.45) *	1.23 (1.02, 1.48) *	1.05 (0.61, 1.79)
UZ	1.83 (1.47, 2.28) ***	1.89 (1.51, 2.37) ***	1.25 (0.49, 3.20)
Males			
Zygosity			
MZ	Ref.	Ref.	Ref.
SSDZ	1.01 (0.80, 1.28)	1.06 (0.83, 1.35)	0.63 (0.29, 1.37)
OSDZ	1.22 (0.97, 1.53)	1.23 (0.97, 1.55)	1.14 (0.56, 2.29)
UZ	1.86 (1.43, 2.42) ***	1.92 (1.46, 2.51) ***	1.26 (0.39, 4.11)
Females			
Zygosity			
MZ	Ref.	Ref.	Ref.
SSDZ	1.17 (0.88, 1.56)	1.19 (0.88, 1.62)	1.04 (0.46, 2.31)
OSDZ	1.19 (0.90, 1.58)	1.23 (0.91, 1.67)	1.14 (0.56, 2.28)
UZ	1.71 (1.13, 2.60) *	1.78 (1.15, 2.75) **	1.18 (0.26, 5.45)

[a] Baseline hazard stratified on sex. * $p < 0.05$, ** $p < 0.01$, *** $p < 0.001$.

There are no interactions between zygosity and age-group among male or female twins (all: $p = 0.696$, males: $p = 0.519$, females: $p = 0.902$). In all three sensitivity analyses, treating all UZ twin pairs as MZ, randomizing UZ twin pairs 1:1 as MZ or SSDZ twin pairs, or randomizing UZ twin pairs 1:2 as MZ or ss-DZ twin pairs, there are no differences in mortality between MZ and DZ twin individuals in neither age group (Supplementary Tables S1–S3).

For MZ and SSDZ twin individuals, the upper-tail of lifespan distributions are similar for both male and female twins. Overall for the birth cohorts 1870–1930, the estimated values of the shape parameters, reflecting the probability of reaching very long lifespans, are very similar for MZ males and DZ males (MZ: ξ (95%-CI) = −0.74 (−0.76, −0.69); DZ: ξ (95%-CI) = −0.75 (−0.77, −0.72)), and for MZ and DZ females (MZ: ξ (95%-CI) = −0.86 (−0.88, −0.81); DZ: ξ (95%-CI) = −0.84 (−0.86, −0.81)). Hence, the shape of the upper-tail of lifespans are very similar by zygosity suggesting the same mortality process regarding the probability of reaching very long lifespans.

4. Discussion

4.1. Main Findings

Using mortality information for the 1870–1990 Danish twin cohorts, we find that the mortality after infancy of MZ and DZ twin individuals differ slightly after taking age, sex, and birth cohort into account. However, when further taking into consideration that some twin pairs have unknown zygosity, we find no indication of a substantial or systematic difference in survival between MZ and DZ twin individuals. Sensitivity analyses, treating UZ twin pairs as MZ twins or randomizing UZ twin pairs as either MZ or DZ twins, indicate that any apparent differences in mortality after infancy between MZ and DZ twin individuals disappear when accounting for UZ twins, suggesting that the differences in mortality between MZ and DZ twins may be explained by selection due to unknown zygosity of some twins. Finally, we find that the distribution of long-livers, that is, the upper-tail of the lifespan distribution, is very similar in MZ and SSDZ twin individuals suggesting the same mortality process regarding the probability of reaching very long lifespans.

4.2. Comparisons with Other Studies

In line with the results of Sharrow and Anderson [9], survival analyses, not accounting for UZ twins, suggest better survival of MZ than DZ twin individuals in the younger age groups. However, these results might reflect a deselection of UZ twins with higher mortality than twins with known zygosity. The differences in mortality after infancy between MZ and DZ twin individuals vanish when UZ twins are merged with twins with known zygosities under different conditions. The analyses in the present paper as well as the paper by Sharrow & Anderson [9] are based on left truncated data, conditioning on both twins surviving to age six (present paper) or 10 (Sharrow and Anderson [9]), and may thus include only the most robust twin pairs where both of them were strong enough to survive their youngest childhood. Including the individual surviving twin from all pairs in which one twin died under the age of six would enhance the comparability with the general population on mortality after infancy. Unfortunately, there are no data to enable follow-up of such surviving twins from the early cohorts.

That MZ and DZ twins derive from the same base population with respect to lifespan is consistent with other literature indicating no or little differences between MZ and DZ twin individuals, and between twins and singletons in adulthood with respect to mortality, lifestyle, and incidence of common diseases [6,22–26]. Data from multiple twin cohorts around the world have formed a substantial part of genome-wide association studies of common traits and complex diseases and there is no evidence to show that the gene–disease associations seen in singletons and twins differ. Thus, being a twin does not appear to impact the basic biological processes and human development in adolescence and adulthood. This implies that findings from twins are generalizable to the population as a whole. Given that twin studies often have response rates higher than surveys in the population at large, twin studies can be considered more representative when based on large, population-derived twin cohorts such as the Danish Twin Registry and other Nordic cohorts.

4.3. Strengths and Weaknesses

The results in this paper are based on a very large dataset comprising more than 90,000 individual twins with information of high quality on zygosity and vital status. The mortality of individual twins is analyzed using an empirical approach and semiparametric models to conduct statistical inference, testing hypotheses and estimating hazard ratios, while accounting for the dependency within twin pairs and adjusting for birth cohorts by stratifying the baseline hazard function.

Although the Danish Twin Registry was established in 1954, it includes birth cohorts from 1870 onwards. For twins in the earlier birth cohorts, before the register was established, zygosity of the twins was determined late in their lives, or even posthumously, leading to a relatively large proportion of twin pairs with UZ in the early cohorts. Due to the structure in church records, retrospective identification of opposite-sex twins was more difficult than identification of same-sex twins, which may explain why some of the earlier birth cohorts in the Danish Twin Registry do not capture as large a proportion of opposite-sex twins as same-sex twins. In the third cohort, birth cohorts 1930–1960, the presence of MZ and female twin individuals are lower than in the other cohorts which might lead to selection bias. To address potential confounding due to this bias in the survival analyses, all baseline hazards were stratified on cohorts and, in analyses combing both sexes, the baseline hazards were further stratified on sex. The criterion on survival until 1968 for all birth cohorts from 1930–1968 could introduce a healthy-survivor effect, although this effect is likely to be comparable for MZ and DZ twin individuals.

In the 19th and first part of the 20th century, infant mortality was high overall and especially high for twins due to low birth weight. In the current paper as well as the paper by Sharrow and Anderson [9], twin pairs were included only if both had survived to age 6, respectively age 10, thus selecting only the hardiest twin pairs. This could lead to biased results on survival in favour of MZ twins over DZ twins because MZ twins have higher infant mortality. Another possible selection bias in the earlier cohorts may have been difficulties in classifying the zygosity of twins dying young since, in many cases, also their parents had died and they could not act as informants about zygosity. Thus, the high mortality seen in UZ twins in the early cohorts might be related to high mortality in the population in general rather than implying high early mortality in UZ twins.

5. Conclusions

The population-based and oldest twin cohort ever studied suggests that although direct comparisons of MZ and DZ twin individuals may indicate that they differ slightly in mortality, no substantial nor systematic difference in survival between MZ and DZ twin individuals is found when taking twins with unknown zygosity into consideration and accounting for the effects of birth- and age-cohorts, gender differences, and that twins are paired.

Supplementary Materials: The following are available online at http://www.mdpi.com/2073-4425/10/2/166/s1, Figure S1: Kaplan-Meier survival curves of mortality for males in each of the four cohorts; Figure S2: Kaplan-Meier survival curves of mortality for females in each of the four cohorts; Figure S3: Kaplan-Meier survival curves for males and females in birth cohorts 1961–1990; Table S1: Age stratified associations between mortality and zygosity in individual twins, treating all UZ-twin pairs as MZ, birth cohorts 1961–1990; Table S2: Age stratified associations between mortality and zygosity in individual twins, randomising UZ-twin pairs 1:1 to either MZ or same-sex DZ, birth cohorts 1961–1990; Table S3: Age stratified associations between mortality and zygosity in individual twins, randomising UZ-twin pairs 1:2 to either MZ or same-sex DZ, birth cohorts 1961–1990.

Author Contributions: Conceptualization J.H. and K.C.; methodology J.H. and P.L.; formal analysis J.H and P.L.; project administration J.H. and P.L.; writing—original draft preparation, J.H. and P.L.; writing—review and editing J.H., P.L., J.K., M.M., T.S., P.H. and K.C.

Funding: J.K. has been supported by the Academy of Finland (grants 308248 & 312073).

Conflicts of Interest: The authors declare no conflict of interest.

References

1. Polderman, T.J.; Benyamin, B.; de Leeuw, C.A.; Sullivan, P.F.; van Bochoven, A.; Visscher, P.M.; Posthuma, D. Meta-analysis of the heritability of human traits based on fifty years of twin studies. *Nat. Genet.* **2015**, *47*, 702–709. [CrossRef] [PubMed]
2. Ahrenfeldt, L.J.; Larsen, L.A.; Lindahl-Jacobsen, R.; Skytthe, A.; Hjelmborg, J.V.; Moller, S.; Christensen, K. Early-life mortality risks in opposite-sex and same-sex twins: A Danish cohort study of the twin testosterone transfer hypothesis. *Ann. Epidemiol.* **2017**, *27*, 115–120. [CrossRef] [PubMed]
3. Loos, R.; Derom, C.; Vlietinck, R.; Derom, R. The East Flanders Prospective Twin Survey (Belgium): A population-based register. *Twin Res.* **1998**, *1*, 167–175. [CrossRef] [PubMed]
4. Hjelmborg, J.V.B.; Iachine, I.; Skytthe, A.; Vaupel, J.W.; McGue, M.; Koskenvuo, M.; Kaprio, J.; Pedersen, N.L.; Christensen, K. Genetic influence on human lifespan and longevity. *Hum. Genet.* **2006**, *119*, 312–321. [CrossRef] [PubMed]
5. Scheike, T.H.; Holst, K.K.; von Bornemann Hjelmborg, J. Estimating twin concordance for bivariate competing risks twin data. *Stat. Med.* **2014**, *33*, 1193–1204. [CrossRef] [PubMed]
6. Christensen, K.; Vaupel, J.W.; Holm, N.V.; Yashin, A.I. Mortality among twins after age 6: Fetal origins hypothesis versus twin method. *BMJ* **1995**, *310*, 432–436. [CrossRef] [PubMed]
7. Kaprio, J. The Finnish Twin Cohort Study: An update. *Twin Res. Hum. Genet.* **2013**, *16*, 157–162. [CrossRef] [PubMed]
8. Kleinman, J.C.; Fowler, M.G.; Kessel, S.S. Comparison of infant mortality among twins and singletons: United States 1960 and 1983. *Am. J. Epidemiol.* **1991**, *133*, 133–143. [CrossRef] [PubMed]
9. Sharrow, D.J.; Anderson, J.J. A Twin Protection Effect? Explaining Twin Survival Advantages with a Two-Process Mortality Model. *PLoS ONE* **2016**, *11*, e0154774. [CrossRef] [PubMed]

10. Skytthe, A.; Kyvik, K.; Holm, N.V.; Vaupel, J.W.; Christensen, K. The Danish Twin Registry: 127 birth cohorts of twins. *Twin Res.* **2002**, *5*, 352–357. [CrossRef] [PubMed]
11. Skytthe, A.; Kyvik, K.O.; Holm, N.V.; Christensen, K. The Danish Twin Registry. *Scand. J. Public Health* **2011**, *39*, 75–78. [CrossRef] [PubMed]
12. Sarna, S.; Kaprio, J.; Sistonen, P.; Koskenvuo, M. Diagnosis of twin zygosity by mailed questionnaire. *Hum. Hered.* **1978**, *28*, 241–254. [CrossRef] [PubMed]
13. Pedersen, C.B.; Gotzsche, H.; Moller, J.O.; Mortensen, P.B. The Danish Civil Registration System. A cohort of eight million persons. *Dan. Med. Bull.* **2006**, *53*, 441–449. [PubMed]
14. Christiansen, L.; Frederiksen, H.; Schousboe, K.; Skytthe, A.; von Wurmb-Schwark, N.; Christensen, K.; Kyvik, K. Age- and sex-differences in the validity of questionnaire-based zygosity in twins. *Twin Res.* **2003**, *6*, 275–278. [CrossRef] [PubMed]
15. Cederlof, R.; Friberg, L.; Jonsson, E.; Kaij, L. Studies on similarity diagnosis in twins with the aid of mailed questionnaires. *Acta Genet. Stat. Med.* **1961**, *11*, 338–362. [CrossRef] [PubMed]
16. Hauge, M.; Harvald, B.; Fischer, M.; Gotlieb-Jensen, K.; Juel-Nielsen, N.; Raebild, I.; Shapiro, R.; Videbech, T. The Danish Twin Register. In *Prospective Longitudinal Research: An Empirical Basis for the Primary Pervention of Psychological Disorders*; Mednich, S.A., Ed.; Oxford University Press: Oxford, UK, 1981; pp. 217–221.
17. Dignam, J.J.; Zhang, Q.; Kocherginsky, M. The use and interpretation of competing risks regression models. *Clin. Cancer Res.* **2012**, *18*, 2301–2308. [CrossRef] [PubMed]
18. Arellano, M. PRACTITIONERS' CORNER: Computing Robust Standard Errors for Within-groups Estimators. *Oxf. Bull. Econ. Stat.* **1987**, *49*, 431–434. [CrossRef]
19. Coles, S. *An Introduction to Statistical Modeling of Extreme Values*; Springer: London, UK, 2001.
20. Medford, A. Best-practice life expectancy: An extreme value approach. *Demogr. Res.* **2017**, *36*, 989–1014. [CrossRef]
21. Gilleland, E.; Katz, R.W. extRemes 2.0: An Extreme Value Analysis Package in R. *J. Stat. Soft.* **2016**, *72*, 1–39. [CrossRef]
22. Christensen, K.; McGue, M. Commentary: Twins, worms and life course epidemiology. *Int. J. Epidemiol.* **2012**, *41*, 1010–1011. [CrossRef] [PubMed]
23. Petersen, I.; Nielsen, M.M.; Beck-Nielsen, H.; Christensen, K. No evidence of a higher 10 year period prevalence of diabetes among 77,885 twins compared with 215,264 singletons from the Danish birth cohorts 1910–1989. *Diabetologia* **2011**, *54*, 2016–2024. [CrossRef] [PubMed]
24. Christensen, K.; Petersen, I.; Skytthe, A.; Herskind, A.M.; McGue, M.; Bingley, P. Comparison of academic performance of twins and singletons in adolescence: Follow-up study. *BMJ* **2006**, *333*, 1095. [CrossRef] [PubMed]
25. Oberg, S.; Cnattingius, S.; Sandin, S.; Lichtenstein, P.; Morley, R.; Iliadou, A.N. Twinship influence on morbidity and mortality across the lifespan. *Int. J. Epidemiol.* **2012**, *41*, 1002–1009. [CrossRef] [PubMed]
26. Ahrenfeldt, L.J.; Skytthe, A.; Moller, S.; Czene, K.; Adami, H.O.; Mucci, L.A.; Kaprio, J.; Petersen, I.; Christensen, K.; Lindahl-Jacobsen, R. Risk of Sex-Specific Cancers in Opposite-Sex and Same-Sex Twins in Denmark and Sweden. *Cancer Epidemiol. Biomarkers Prev.* **2015**, *24*, 1622–1628. [CrossRef] [PubMed]

GCAT
TACG
GCAT
genes

MDPI

Article

Analysis of the Association Between *TERC* and *TERT* Genetic Variation and Leukocyte Telomere Length and Human Lifespan—A Follow-Up Study

Daniela Scarabino [1], Martina Peconi [2], Franca Pelliccia [3] and Rosa Maria Corbo [1,3,*]

[1] CNR Institute of Molecular Biology and Pathology, P.le Aldo Moro 5, 00185 Rome, Italy;
    daniela.scarabino@cnr.it
[2] CNR Institute of Translational Pharmacology, Via Fosso del Cavaliere 100, 00133 Rome, Italy;
    martinapeconi@gmail.com
[3] Department of Biology and Biotechnology, La Sapienza University, P.le Aldo Moro 5, 00185 Rome, Italy;
    franca.pelliccia@uniroma1.it
*   Correspondence: rosamaria.corbo@uniroma1.it

Received: 17 December 2018; Accepted: 23 January 2019; Published: 25 January 2019

Abstract: We investigated the possible influence of *TERC* and *TERT* genetic variation and leukocyte telomere length (LTL) on human lifespan. Four polymorphisms of *TERT* and three polymorphisms of *TERC* were examined in a sample of elderly subjects (70–100 years). After nine years of follow-up, mortality data were collected, and sub-samples of long-lived/not long-lived were defined. *TERT* VNTR MNS16A L/L genotype and *TERT* rs2853691 A/G or G/G genotypes were found to be associated with a significantly higher risk to die before the age of 90 years, and with a significantly lower age at death. The association between lifespan and LTL at baseline was analyzed in a subsample of 163 subjects. Age at baseline was inversely associated with LTL ($p < 0.0001$). Mean LTL was greater in the subjects still living than in those no longer living at follow-up (0.79 T/S $\pm$ 0.09 vs. 0.63 T/S $\pm$ 0.08, $p < 0.0001$). Comparison of age classes showed that, among the 70–79-year-olds, the difference in mean LTL between those still living and those no longer living at follow-up was greater than among the 80–90-year-olds. Our data provide evidence that shorter LTL at baseline may predict a shorter lifespan, but the reliability of LTL as a lifespan biomarker seems to be limited to a specific age (70–79 years).

Keywords: human lifespan; genetic variation; *TERC*; *TERT*; leukocyte telomere length

1. Introduction

The dramatic increase in rates of survival to an advanced old age over the past century has prompted extensive research in the attempt to identify the mechanisms involved in lifespan determination. Among the most extensively studied biological processes associated with longevity are those involved in cell maintenance/senescence. Telomeres, the structures at the ends of eukaryotic chromosomes with a protective action against genome instability, have been widely studied as a possible determinant of lifespan [1]. Human telomeres are composed of repeated TTAGGG nucleotide sequences located at the ends of each chromosome. Because telomere sequences are not fully replicated during DNA replication due to the inability of DNA polymerase to replicate the 3' end of the DNA strand, telomeres shorten as cells divide. In the absence of special telomere maintenance mechanisms, telomeres (and chromosomes) become shorter with each cell division. Once a critically short telomere length is reached, the cell is triggered to enter replicative senescence, ultimately leading to cell death. Telomerase, a cellular ribonucleoprotein enzyme complex, counteracts telomere shortening [2]. Human telomerase is constituted by a DNA reverse transcriptase polymerase (telomerase reverse transcriptase,

TERT), which uses an RNA template (telomerase RNA component, *TERC*) to add telomeric DNA onto telomeres, thus compensating for the telomere shortening caused by cell divisions [3]. The two components of human telomerase are encoded by the *TERT* gene on 5p15.33 (OMIM:187270) and by the *TERC* gene on 3q26 (OMIM:602322). Since telomerase is almost totally absent in adult tissues, including the skin, kidney, liver, blood vessels, and peripheral leukocytes, the telomeres of replicating cells shorten progressively. This mechanism is thought to underlie aging and age-associated diseases [4–6]. Average leukocyte telomere length (LTL) is generally used as a marker of overall telomere length, since TLs have been found to be strongly correlated across different cell types within the same individual [7,8]. Population studies that have applied analysis of LTL support the hypothesis that leucocyte telomere shortening is associated with aging and lifespan [4–6,9,10]; however, the associations with age-related chronic diseases (cardiovascular and metabolic disease, cancer) are not always concordant [8,9,11,12].

As fully functional telomerase is critical for telomere maintenance, genetic variations of human *TERT* and *TERC* genes may alter the stability of the telomerase complex or directly affect its enzymatic activity [13]. Studies assessing the possible effect of genetic polymorphisms of human *TERT* and *TERC* genes on LTL [13–16], and on aging and lifespan [17–19], have produced mixed results. While some *TERC* or *TERT* SNPs were found to be associated with longevity, the relation was not always mediated by the association with telomere length. Similar contradictory results have come from genetic association studies of *TERT* polymorphisms and common diseases [8].

In the present study, we investigated the possible impact on the human lifespan of four polymorphisms of the *TERT* gene (MNS16A, rs2853691, rs33954691, rs2736098) and three polymorphisms of the *TERC* gene (rs12696304, rs3772190, rs16847897). MNS16A is a minisatellite (variable number of tandem repeats, VNTR) located downstream of exon 16 of the *TERT* gene and upstream in the putative promoter region of an antisense *TERT* transcript. It shows two common alleles (VNTR-302 or L and VNTR-243 or S on the basis of the PCR fragment size) [20]. It has been studied in relation to longevity and cancer risk [18,21–23]. The detection of antisense *TERT* mRNA suggested its possible role in regulating human telomerase expression [20]. *TERT* rs2853691 is located in an intronic region and shows two common alleles, A and G, while rs33954691 is located in exon 14, where a C to T substitution does not result in a change of the amino acid (Histidine) at codon 1013. These two *TERT* SNPs have been reported to be associated with both LTL and lifespan [13,17]. rs2736098 is located in exon 2, where a G to A substitution does not result in a change of the amino acid alanine at codon 305; it has shown a strong association with some cancer types (see OMIM %613059). The *TERC* SNPs rs12696304, rs3772190, and rs16847897 are all located downstream of *TERC* in a noncoding region [14,17] and have been consistently associated with variation of LTL [14,16,17,24]. In addition, in the attempt to gain a better understanding of the relationships between telomere length and lifespan, in the present study, we analyzed LTL in a subsample of elderly subjects who had been genotyped for *TERT* and *TERC* polymorphisms.

The association between LTL and *TERT* and *TERC* polymorphisms and longevity was investigated by means of a follow-up study. The study sample was originally recruited in 2000. After collecting mortality information in 2009, we defined a sample of long-lived subjects as those who died after the age of 90 years, and a sample of not-long-lived subjects composed of those who had died before reaching 90 years of age.

2. Materials and Methods

2.1. Materials

The sample was recruited in 2000 for the multidisciplinary LONCILE (Longevity of Cilento) study on the anthropological and biological characteristics of the elderly population of the Cilento area in the district of Salerno, southern Italy [25]. As previously reported [26], it consisted of 277 unrelated individuals (43.7% males) born between 1900 and 1930 (mean age, 82.9 ± 5.7 years ± standard deviation [SD]), enrolled without selection criteria, except age (>70 years) and birth place; they had no manifest pathologies and were healthy, consistent with age. Mortality data on 267 subjects were collected in 2009.

In 2000, 14.5% were aged 90 years old or older. During the nine-year follow-up period, the mortality rate was 62.5% (51.5% men), and 44.9% of the subjects died after the age of 90 years, including those aged 90 at baseline. As the mean life expectancy in this geographic area in 2000 for subjects 83 years old was seven years for women and six years for men (ISTAT, http://demo.istat.it/unitav/index.html), we defined as long-lived those subjects who, at follow-up in 2009, had died at an age of more than 90 years ($\geq$90 years). The sample of the long-lived (*n* = 75) comprised 100% of subjects aged 90 years or older in 2000 and 36.3% of those aged over 80 years in 2000. The sample of the not long-lived (*n* = 89) was made up of individuals who had died between 2000 and 2009 before reaching the age of 90 years.

The protocol for the collection of biological material for the scientific studies was approved by the institutional committees (Local Health Unit, Salerno 3). The study was approved by the Department Board (12/06/2009 session) of the former Department of Genetics and Molecular Biology of La Sapienza University, Rome. Written, informed consent was obtained from all subjects.

2.2. Laboratory Methods

Genomic DNA was extracted according to the salting out procedure described by Miller et al. [27] from venous blood drawn in EDTANa2 as anticoagulant from all subjects after overnight fasting.

TERT VNTR MNS16A was genotyped according to the allelic-specific PCR method, as previously reported [20,23]. Genotyping revealed, in addition to the most common alleles corresponding to 243 bp band and 302 bp band, less frequent but still polymorphic alleles corresponding to 274 bp band and 333 bp band. The genotypes were then classified according to Wang et al. [20]: short allele (S) corresponds to 243 and 274 bp bands and long allele (L) to 302 and 333 bp bands. The MNS16A genotypes were L/L, L/S, and S/S. *TERT* SNPs (rs2853691 and rs33954691) were investigated by polymerase chain reaction amplification followed by restriction fragment length polymorphism (PCR-RFLP), as previously reported [23]. Genotyping of the *TERT* SNP (rs2736098) and the *TERC* SNPs (rs12696304, rs3772190, and rs16847897) was carried out by allelic discrimination using predesigned TaqMan SNP genotyping assays (Applied Biosystems), as previously reported [23]. The genotyping techniques are reported in detail in Supplementary Materials (Figures S1–S3).

The average (of triplicate) telomere length in leukocytes was measured by real-time PCR quantitative analysis (qPCR) on a 7300 real-time PCR instrument (Applied Biosystems). This method allows the determination of the number of copies of telomeric repeats (T) compared to a single copy gene (S) used as a quantitative control (T/S ratio) [28]. The telomere and single-copy gene β-globin (HGB) were analyzed on the same plate in order to reduce inter-assay variability. DNA (35 ng) was amplified in a total volume of 20 µl containing 10 µl of SYBER Select Master Mix (Applied Biosystems); primers for telomeres and the single-copy gene were added to final concentrations of 0.1 µM (Tel Fw), 0.9 µM (Tel Rev), 0.3 µM (HGB Fw), and 0.7 µM (HGB Rev), respectively. The primer sequences were: Tel Fw 5′-CGGTTTGTTTGGGTTTGGGTTTGGGTTTGGGTTTGGGTT-3′; Tel Rev 5′-GGCTTGCCTTACCCTTACCCTTACCCTTACCCTTACCCT-3′; HGB Fw 5′-GCTTCTGACACAACTGTGTTCACTAGCAAC-3′; and HGB Rev 5′-CACCACCAACTTCATCCACGTTCACCTTGC-3′ [29]. The enzyme was activated at 95 °C for 10 min, followed by 40 cycles at 95 °C for 15 s and 60 °C for 1 min. In addition, two standard curves (one for HGB and one for telomere reactions), were prepared for each plate using a reference DNA sample (Control Genomic Human DNA, Applied Biosystems) diluted in series (dilution factor = 2) in order to produce five concentrations of DNA ranging from 50 to 6.25 ng in 20 µL. Measurements were performed in triplicate and are reported as the T/S ratio relative to the calibrator sample to allow for comparison across runs.

2.3. Statistical Analysis

Allelic frequencies were determined by the gene-counting method. Agreement between the observed genotype distributions and those expected according to the Hardy-Weinberg equilibrium was verified with a chi square test. Linkage disequilibrium (LD) between the *TERT* and *TERC* SNPs and haplotype frequencies were estimated by the maximum likelihood method using the EH program (http://www.genemapping.cn/eh.htm) [30]. The differences in allele, genotype, and haplotype frequencies between patients and controls were analyzed with a chi square test. The probability of living to an age over 90 years ($\geq$90 years) or not associated with *TERT* genotypes was estimated by odds ratios (ORs) adjusted for other variables calculated by logistic regression.

Parametric (ANOVA) and non-parametric (Kruskal-Wallis) tests were used to compare the distribution of LTL across long-lived and not long-lived subjects, and the distribution of the mean T/S ratio across the various *TERT* and *TERC* genotypes. Level of significance was set at $p < 0.05$. The relationship between T/S ratio and age was evaluated by regression analysis.

3. Results

To evaluate the involvement of the *TERT* and *TERC* polymorphisms in lifespan determination, genotype frequencies of *TERT* and *TERC* SNPs observed in the long-lived subjects were compared against those observed in the subjects who had died before reaching the age of 90 years (not long-lived) (Table 1). In both groups, the genotype frequencies of *TERT* and *TERC* polymorphisms agreed with those expected according to Hardy-Weinberg equilibrium. No difference in the distribution of *TERC* SNPs and *TERT* SNPs rs33954691 and rs2736098 genotypes was observed between the long-lived and the not long-lived (Table 1). By contrast, a significant defect of the *TERT* VNTR MNS16A L/L genotype ($p = 0.018$), and rs2853691 A/G and G/G genotypes ($p = 0.01$), was found in the long-lived compared to the not long-lived. The two *TERT* polymorphisms were found in strict linkage disequilibrium ($p < 0.0001$, D = 80% of Dmax), with a trend of the MNS16A L allele to be associated with the rs2853691 G allele and the MNS16A S allele with the rs2853691 A allele.

Table 1. *TERT* and *TERC* genotype distribution in long-lived and not long-lived. Percentage is given in brackets.

Gene/Genotype	Not Long-Lived	Long-Lived
TERC rs12696304		
G/G	11 (12.4)	5 (7.0)
G/C	35 (39.3)	31 (43.7)
C/C	43 (48.3)	35 (49.3)
TOTAL	89	71
p		0.52
TERC rs3772190		
C/C	56 (67.5)	40 (58.8)
C/T	22 (26.5)	25 (36.8)
T/T	5 (6.0)	3 (4.4)
TOTAL	83	68
p		0.40
TERC rs16847897		
C/C	7 (8.6)	8 (11.9)
C/G	36 (44.4)	34 (50.7)
G/G	38 (46.9)	25 (37.3)
TOTAL	81	67
p		0.47
TERT VNTR MNS16A		
L/L	32 (36.0)	13 (18.8)
L/S [1]	42 (47.2)	35 (50.7)
S/S [1]	15 (16.9)	21 (30.4)
TOT	89	69
p		0.018

Table 1. *Cont.*

Gene/Genotype	Not Long-Lived	Long-Lived
TERT rs2853691		
A/A	41 (45.1)	48 (67.6)
A/G[1]	42 (46.2)	22 (31.0)
G/G[1]	8 (8.8)	1 (1.4)
TOT	91	71
p	0. 004	
TERT rs33954691		
C/C	77 (85.6)	62 (84.9)
C/T[1]	11 (12.2)	9 (12.3)
T/T[1]	2 (2.2)	2 (2.7)
TOTAL	90	73
p	0.91	
TERT rs2736098		
C/C	56 (64.4)	52 (72.2)
C/T[1]	30 (34.5)	17 (23.6)
T/T[1]	1 (1.1)	3 (4.2)
TOTAL	87	72
p	0. 29	

[1] These genotypes were pooled for the analysis.

Logistic regression analysis was then applied to correctly evaluate the effect of *TERT* genotypes on longevity. In the analysis, the independent variable was the genotype constituted by the combination of MNS16A L/L and rs2853691 A/G or G/G genotypes. The dependent variable was having lived to an age of over 90 years ($\geq$90 years) or not. The results showed that, after adjusting for sex, carrying MNS16A L/L and rs2853691 A/G or G/G genotypes was associated with a significantly lower probability of living to more than 90 years of age (odds ratio [OR] 0.34, 95% confidence interval [CI] 0.15–0.79, p = 0.012), or, in other words, a risk of 2.94 (1/0.34) to die before the age of 90 years. Analysis of the association between *TERT* MNS16A and rs2853691 genotypes and age at death supported previous findings, showing that the L/L genotype and carrying G alleles are associated with a lifespan of less than 90 years (Table 2).

Table 2. Relationship between *TERT* genotypes and age at death (mean $\pm$ SD). In brackets is the number of subjects.

SNP/Genotypes	Age at Death
VNTR MNS16A	
L/L	87.6 $\pm$ 6.0 (45)
L/S	88.4 $\pm$ 5.4 (77)
S/S	90.8 $\pm$ 6.2 (36)
p	0.04
rs2853691	
A/A	89.9 $\pm$ 5.7 (89)
A/G	87.6 $\pm$ 5.7 (64)
G/G	85.8 $\pm$ 3.7 (9)
p	0.01

Leukocyte telomere length (LTL), expressed as the T/S ratio, was measured in a subgroup of 153 subjects at baseline. The mean LTL value at baseline was 0.69 $\pm$ 0.12 T/S (range, 0.49–1.03, median 0.69), with only a slight difference between males and females (males: n = 59, LTL = 0.67 $\pm$ 0.11; females: n = 94, LTL = 0.70 $\pm$ 0.12, p = 0.18). Age at baseline was inversely associated with telomere length. Linear regression ($y = -0.009x + 1.4$, $p < 0.0001$, $n = 153$) (Figure 1) yielded an estimated telomere loss rate of about 0.010 T/S ratio/year.

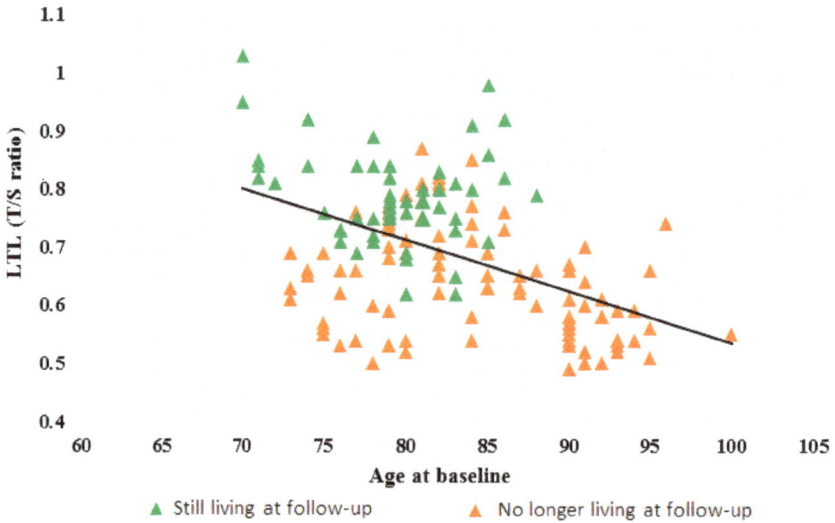

Figure 1. LTL expressed as T/S ratio as a function of age at baseline.

Table 3 lists the estimated haplotype frequencies in the long-lived and the controls. In accordance with single polymorphism observations, a significant defect of the MNS16A L—rs2853691 G haplotype was observed in the long-lived compared to the not long-lived to controls ($p = 0.03$), suggesting that the presence of the two *TERT* alleles may prevent attainment of the oldest ages.

Table 3. *TERT* VNTR MNS16A and *TERT* rs2853691 haplotype distribution in long-lived and not long-lived.

VNTR MNS16A/rs2853691 Haplotype	Not Long-Lived	Long-Lived
L-A	0.294	0.278
L-G	0.301	0.164
S-A [1]	0.397	0.548
S-G [1]	0.008	0.010
p		0.03

[1] These genotypes were pooled for the analysis.

The relationship between LTL at baseline and years of life remaining was analyzed using the follow-up data on lifespan. A significant positive relation was observed (regression line $y = 0.009x + 0.6$, $p = 0.01$, $n = 99$), where the regression coefficient 0.009 T/S provides an estimate of how much longer the telomeres are at baseline for each additional year of life remaining. We then compared the baseline LTL values of the subjects still living at follow-up with those no longer living in both the total sample, and when the sample was divided into three age classes at baseline (70–79, 80–89, and ≥90 years). The mean LTL values at baseline were significantly higher in those still living than in those who had died during the follow-up years. Within each age group, the LTL of those still living was higher than the mean class value and the LTL of those who had died was lower (Table 4). The difference between those who were still living and those who had died was greater among the 70–79-year-olds than among the 80–89-year-olds (Table 4 and Figure 1). The over 90-year-olds had all died during the follow-up period.

Table 4. Mean LTL (T/S ratio) in the total sample, still living and no longer living at follow-up, by age class (mean ± SD).

	All Ages	Age 70–79 Years	Age 80–89 Years	≥ 90 Years
Total sample	0.69 ± 0.12 (153)	0.72 ± 0.12 (54)	0.73 ± 0.10 (63)	0.57 ± 0.06 (36)
No longer living at follow-up	0.63 ± 0.09 (99)	0.63 ± 0.08 (26)	0.69 ± 0.09 (37)	0.57 ± 0.06 (37)
Still living at follow-up	0.79 ± 0.08 (54)	0.80 ± 0.08 (28)	0.77 ± 0.09 (26)	/
p [1]	<0.0001	<0.0001	0.0007	

[1] The p value refers to the comparison between No longer living and Still living at follow-up.

These findings are illustrated in Figure 1: the vast majority of the deceased in the age range 70–79 years had LTL values below the regression line at baseline, whereas those still living had LTL values above the line. Differently, in the higher age range of 80–90 years, the baseline LTL values of the no–longer living and the still living were fairly mixed below/above the line. We then compared the LTL values in the not long-lived (0.67 ± 0.09, n = 56) and the long-lived (0.59 ± 0.08, n = 43, p = 0.003). The long-lived sample, 84% of which were already 90 years old at baseline, had a lower mean LTL value than the not-long-lived, who belonged to younger age groups.

Finally, the mean LTL associated with *TERT* VNTR MNS16A and rs2853691 genotypes involved with lifespan determination was examined in not long-lived and long-lived subjects. No difference in mean LTL was observed among *TERT* VNTR MNS16A and rs2853691 genotypes. However, in subjects carrying the combined risk genotypes MNS16A L/L + rs2853691 G/G or A/G (Table 5), LTL was found to be significantly shorter in the Not long-lived than in the Long-lived (p = 0.05). No difference in mean LTL was observed among the genotypes of the other *TERC* and *TERT* SNPs (data not reported).

Table 5. Mean LTL (T/S ratio) associated with the combined genotypes of MNS16A/rs2853691 polymorphisms.

	L/L + G/G or A/G	L/S or S/S + A/A
Not long-lived	0.63 ± 0.08 (16)	0.64 ± 0.11 (40)
Long-lived	0.72 ± 0.10 (6)	0.66 ± 0.06 (37)
p	0.05	0.57

4. Discussion

Here, we investigated a possible association between *TERT* and *TERC* polymorphisms and LTL and lifespan by means of a follow-up study. This study design allowed us to extend the investigation to include a sample for which the lifespan was known, and to distinguish between a sample of subjects definitely not long-lived and a sample of long-lived subjects. Furthermore, all the subjects belonged to the same birth cohort and had experienced similar social and environmental influences.

Examination of the genetic variation of *TERT* and *TERC* genes showed a significant association between two *TERT* polymorphisms (the minisatellite MNS16A and the SNP rs2853691) and lifespan. Carrying the *TERT* VNTR MNS16A L/L genotype and rs2853691 A/G and G/G genotypes turned out to be associated with an increased risk (2.94) of dying before the age of 90 years, i.e., below the mean life expectancy for subjects living in this geographic area, with an average age of about 83 years at study baseline. The observation was confirmed "in vivo" by mortality data and showed that the same risk genotypes were associated with the shortest lifespan. An association between VNTR MNS16A genotypes and longevity has been observed by Concetti et al. [18], whereas rs2853691 SNP has been reported to belong to a haplotype involved in longevity [13,17]. Our data support these findings and highlight that the MNS16A L/L genotype and rs2853691 A/G and G/G prevent the attainment of longevity. Consistent with this result are previous findings that the MNS16A L/L genotype is associated with an increased risk of Alzheimer's disease [23] and lower survival in patients with glioblastoma or lung cancer [31,32], and that rs2853691 A/G and G/G are associated with esophageal squamous cell carcinoma [33]. The shorter lifespan associated with *TERT* genotypes would therefore, at least in part, be explained by their involvement in the onset of aging-related diseases. We observed a marginally significant association of the combined risk genotypes of the two *TERT* polymorphisms with shorter LTL in the Not long-lived subjects. Although the sample was quite small, this observation is consistent with a previous work [18] that reported a tendency of greater telomere shortening in elderly subjects with homozygous VNTR MNS16A L/L as compared with the other MNS16A genotypes. Considering that the L allele seems to have a negative regulatory role in the expression of telomerase [20], the overall picture suggests that the relationship of *TERT* genotypes with lifespan is mediated by an action of *TERT* on telomere length. We found no significant relationships of the *TERC* SNPs with longevity or telomere length, although the *TERC* SNPs we examined were often found to be associated with telomere length [14,16,17,19], and with longevity (rs3772190) [17]. The conflicting data might depend on diverse factors such as sample size, population examined, and mean age of the population sample, among others. Telomere length being a complex character, numerous genes will contribute to its determination, each providing a small contribution that could be difficult to distinguish. In addition, the interaction of genetic determinants with environmental factors, such as different population lifestyles, could explain the inconsistencies. Furthermore, the discordant results might depend on the technique used for LTL measurement as well. In the majority of the population studies cited above, the average length of telomeres was measured by Real-Time PCR quantitative analysis (qPCR) or by Southern blot analysis of the terminal restriction fragments, and there is evidence that intra- and inter-laboratory technical variation severely limits the comparability of telomere length estimates between laboratories [34].

Here, we also examined the relationships between LTL and lifespan. A significant negative correlation between age and LTL at baseline was observed, with an estimated telomere loss rate of 0.009 T/S ratio/year. This observation is shared by previous studies [35]; the yearly telomere loss was very similar to the reported value (0.010 T/S ratio/year) [35]. The mortality data provided by the follow-up allowed us to evaluate the relationship between LTL at baseline and the number of the remaining years of life. The positive relationship we observed indicates that the shorter the telomeres at baseline, the fewer the remaining years of life. In line with this result, analysis of the mean LTL values at baseline showed that the mean LTL was much shorter in those who had died within nine years of follow-up than in those still alive at follow-up (Table 4). This is partly due to the fact that among the deceased, all were already 90 years old at baseline (about 36%), and therefore with reduced telomeres according to age. The remaining 64% included subjects who had died before the age of 90 and who, as can be seen from the LTL data in Table 4, had LTL values lower than both the average value of their age class and the average value of those still living at follow-up. This is illustrated in Figure 1, where the LTL values for subjects who died during the follow-up are mostly distributed below the regression line. Taking into account the different age classes (Table 4), a comparison of the mean LTL between those still living and those no longer living at follow-up, showed a greater difference in the 70-to-79-year olds compared with the 80-to-90-year-olds, in which the LTL of the no longer living was closer to the LTL of the still living. Again, this pattern is clearly shown in Figure 1: before the age of 80 years, the LTL values of the still living and the no longer living are well-separated by the regression line, but they become quite mixed after the age of 80 years. On the whole the picture provided by our data indicates that telomere length is related to lifespan. Indeed, it is almost a lifespan biomarker. However, this relationship is stronger for the younger age group (70–79), and then weakens after 80 years of age.

Despite the mixed results [36] of epidemiologic studies investigating the link between LTL and lifespan/mortality, accumulating data tend to confirm that shorter baseline TL is a marker of greater susceptibility to age-related diseases and of higher overall mortality risk [37–39]. Varying sample sizes and other characteristics, such as age range or the length of the follow-up period, underlie the conflicting data. In addition, the wide inter-individual variability in telomere length for individuals of the same chronological age due to inherited and environmental factors may mask any relationships between LTL and lifespan [40].

The association between TL and lifespan we observed seemed to weaken in the older age classes. Several studies have reported that the magnitude of the association of shorter LTL with higher mortality rates declines with increasing age [4,38–43]. A plausible explanation is the so-called "survival bias". In collecting study samples of older individuals, subjects with shorter baseline TL, being more susceptible to age-related diseases, may be less likely to be included in the study [36,37]. This would lead to a reduced variability of LTL measurements, shifted towards a longer telomere length. Furthermore, leukocyte telomere shortening reflects active cell proliferation triggered by factors such as oxidative stress and chronic systemic inflammation, both of which are aging-related processes. The overexpression of proinflammatory cytokines and mediators observed in older individuals, activating leukocyte proliferation, may lead to alterations in the relationship between LTL and age, and ultimately lifespan [38,44,45]. In this context, it could be hypothesized that, at an age before 80 years, shorter telomere length may indicate a greater susceptibility to aging-related diseases and therefore be predictive of reduced lifespan, whereas in those older than 80 years of age, chronic systemic inflammation, together with oxidative stress, could act as prevailing determinants of leukocyte proliferation and telomere shortening, thus making the relationship between telomere length and lifespan less linear. LTL therefore seems to be a more reliable lifespan biomarker in a younger age class.

There are indications that the relationship between LTL and age depends on nongenetic factors such as age, sex, race/ethnicity, lifestyle practices, and dietary patterns [10]. The strength of the present paper is the examination of a fairly well-defined population sample for ethnicity, age range, lifestyle, and dietary patterns. This homogeneity allowed us to define a temporal window, the interval of 70–79 years, in which LTL could be seen as a good lifespan biomarker. These observations provide useful indications for designing investigations that aim to assess the possible use of LTL as a lifespan predictor.

Supplementary Materials: The following are available online at http://www.mdpi.com/2073-4425/10/2/82/s1, Figure S1: Genotyping of TERT MNS16A; Figure S2: Genotyping of TERT rs2853691; Figure S3: Genotyping of TERT rs33954691.

Author Contributions: Conceptualization, D.S. and R.M.C.; investigation, D.S., M.P. and F.P.; data curation: M.P.; formal analysis and writing, R.M.C. and D.S.; funding acquisition, F.P. and R.M.C. All authors read and approve the paper.

Funding: This work was supported by La Sapienza University grants 2016/2017.

Acknowledgments: We wish to thank K.A. Britsch for checking the manuscript for style.

Conflicts of Interest: The authors have no competing interests to declare.

References

1. Blackburn, E.H.; Epel, E.S.; Lin, J. Human telomere biology: A contributory and interactive factor in aging, disease risks, and protection. *Science* **2015**, *350*, 1193–1198. [CrossRef] [PubMed]
2. Wong, J.M.Y.; Collins, K. Telomere maintenance and disease. *Lancet* **2003**, *362*, 983–988. [CrossRef]
3. Blackburn, E.H.; Greider, C.W.; Szostak, J.W. Telomeres and telomerase: The path from maize, Tetrahymena and yeast to human cancer and aging. *Nat. Med.* **2006**, *12*, 1133–1138. [CrossRef] [PubMed]
4. Cawthon, R.M.; Smith, K.R.; O'Brien, E.; Sivatchenko, A.; Kerber, R.A. Association between telomere length in blood and mortality in people aged 60 years or older. *Lancet* **2003**, *361*, 393–395. [CrossRef]
5. Kimura, M.; Hjelmborg, J.V.B.; Gardner, J.P.; Bathum, L.; Brimacombe, M.; Lu, X.; Christiansen, L.; Vaupel, J.W.; Aviv, A.; Christensen, K. Telomere length and mortality: A study of leukocytes in elderly Danish twins. *Am. J. Epidemiol.* **2008**, *167*, 799–806. [CrossRef] [PubMed]
6. Lapham, K.; Kvale, M.N.; Lin, J.; Connell, S.; Croen, L.A.; Dispensa, B.P.; Fang, L.; Hesselson, S.; Hoffmann, T.J.; Iribarren, C.; et al. Automated Assay of Telomere Length Measurement and Informatics for 100,000 Subjects in the Genetic Epidemiology Research on Adult Health and Aging (GERA) Cohort. *Genetics* **2015**, *200*, 1061–1072. [CrossRef] [PubMed]
7. Daniali, L.; Benetos, A.; Susser, E.; Kark, J.D.; Labat, C.; Kimura, M.; Desai, K.; Granick, M.; Aviv, A. Telomeres shorten at equivalent rates in somatic tissues of adults. *Nat. Commun.* **2013**, *4*, 1597. [CrossRef]

8. Barrett, J.H.; Iles, M.M.; Dunning, A.M.; Pooley, K.A. Telomere length and common disease: Study design and analytical challenges. *Hum. Genet.* **2015**, *134*, 679–689. [CrossRef]
9. Sanders, J.L.; Newman, A.B. Telomere length in epidemiology: A biomarker of aging, age-related disease, both, or neither? *Epidemiol. Rev.* **2013**, *35*, 112–131. [CrossRef]
10. Needham, B.L.; Rehkopf, D.; Adler, N.; Gregorich, S.; Lin, J.; Blackburn, E.H.; Epel, E.S. Leukocyte telomere length and mortality in the National Health and Nutrition Examination Survey, 1999–2002. *Epidemiology* **2015**, *26*, 528–535. [CrossRef]
11. Codd, V.; Nelson, C.P.; Albrecht, E.; Mangino, M.; Deelen, J.; Buxton, J.L.; Hottenga, J.J.; Fischer, K.; Esko, T.; Surakka, I.; et al. Identification of seven loci affecting mean telomere length and their association with disease. *Nat. Genet.* **2013**, *45*, 422–427. [CrossRef] [PubMed]
12. Scheller Madrid, A.; Rode, L.; Nordestgaard, B.G.; Bojesen, S.E. Short Telomere Length and Ischemic Heart Disease: Observational and Genetic Studies in 290 022 Individuals. *Clin. Chem.* **2016**, *62*, 1140–1149. [CrossRef] [PubMed]
13. Atzmon, G.; Cho, M.; Cawthon, R.M.; Budagov, T.; Katz, M.; Yang, X.; Siegel, G.; Bergman, A.; Huffman, D.M.; Schechter, C.B.; et al. Evolution in health and medicine Sackler colloquium: Genetic variation in human telomerase is associated with telomere length in Ashkenazi centenarians. *Proc. Natl. Acad. Sci. USA* **2010**, *107* (Suppl. 1), 1710–1717. [CrossRef] [PubMed]
14. Codd, V.; Mangino, M.; van der Harst, P.; Braund, P.S.; Kaiser, M.; Beveridge, A.J.; Rafelt, S.; Moore, J.; Nelson, C.; Soranzo, N.; et al. Common variants near *TERC* are associated with mean telomere length. *Nat. Genet.* **2010**, *42*, 197–199. [CrossRef] [PubMed]
15. Njajou, O.T.; Blackburn, E.H.; Pawlikowska, L.; Mangino, M.; Damcott, C.M.; Kwok, P.-Y.; Spector, T.D.; Newman, A.B.; Harris, T.B.; Cummings, S.R.; et al. A common variant in the telomerase RNA component is associated with short telomere length. *PLoS ONE* **2010**, *5*, e13048. [CrossRef]
16. Shen, Q.; Zhang, Z.; Yu, L.; Cao, L.; Zhou, D.; Kan, M.; Li, B.; Zhang, D.; He, L.; Liu, Y. Common variants near *TERC* are associated with leukocyte telomere length in the Chinese Han population. *Eur. J. Hum. Genet.* **2011**, *19*, 721–723. [CrossRef]
17. Soerensen, M.; Thinggaard, M.; Nygaard, M.; Dato, S.; Tan, Q.; Hjelmborg, J.; Andersen-Ranberg, K.; Stevnsner, T.; Bohr, V.A.; Kimura, M.; et al. Genetic variation in *TERT* and *TERC* and human leukocyte telomere length and longevity: A cross-sectional and longitudinal analysis. *Aging Cell* **2012**, *11*, 223–227. [CrossRef]
18. Concetti, F.; Lucarini, N.; Carpi, F.M.; Di Pietro, F.; Dato, S.; Capitani, M.; Nabissi, M.; Santoni, G.; Mignini, F.; Passarino, G.; et al. The functional VNTR MNS16A of the *TERT* gene is associated with human longevity in a population of Central Italy. *Exp. Gerontol.* **2013**, *48*, 587–592. [CrossRef]
19. Crocco, P.; Barale, R.; Rose, G.; Rizzato, C.; Santoro, A.; De Rango, F.; Carrai, M.; Fogar, P.; Monti, D.; Biondi, F.; et al. Population-specific association of genes for telomere-associated proteins with longevity in an Italian population. *Biogerontology* **2015**, *16*, 353–364. [CrossRef]
20. Wang, L.; Soria, J.-C.; Chang, Y.-S.; Lee, H.-Y.; Wei, Q.; Mao, L. Association of a functional tandem repeats in the downstream of human telomerase gene and lung cancer. *Oncogene* **2003**, *22*, 7123–7129. [CrossRef]
21. Hofer, P.; Baierl, A.; Feik, E.; Führlinger, G.; Leeb, G.; Mach, K.; Holzmann, K.; Micksche, M.; Gsur, A. MNS16A tandem repeats minisatellite of human telomerase gene: A risk factor for colorectal cancer. *Carcinogenesis* **2011**, *32*, 866–871. [CrossRef] [PubMed]
22. Hashemi, M.; Amininia, S.; Ebrahimi, M.; Hashemi, S.M.; Taheri, M.; Ghavami, S. Association between h*TERT* polymorphisms and the risk of breast cancer in a sample of Southeast Iranian population. *BMC Res. Notes* **2014**, *7*, 895. [CrossRef] [PubMed]
23. Scarabino, D.; Broggio, E.; Gambina, G.; Pelliccia, F.; Corbo, R.M. Common variants of human *TERT* and *TERC* genes and susceptibility to sporadic Alzheimers disease. *Exp. Gerontol.* **2017**, *88*, 19–24. [CrossRef] [PubMed]
24. Maubaret, C.G.; Salpea, K.D.; Romanoski, C.E.; Folkersen, L.; Cooper, J.A.; Stephanou, C.; Li, K.W.; Palmen, J.; Hamsten, A.; Neil, A.; et al. Association of *TERC* and *OBFC1* haplotypes with mean leukocyte telomere length and risk for coronary heart disease. *PLoS ONE* **2013**, *8*, e83122. [CrossRef] [PubMed]
25. Cresta, M.; Gregorio, G. Il disegno dello studio LONCILE. *Riv. Antropol.* **2001**, *79*, 11–18.
26. Corbo, R.M.; Pinto, A.; Scacchi, R. Gender-specific association between FSHR and PPARG common variants and human longevity. *Rejuvenation Res.* **2013**, *16*, 21–27. [CrossRef] [PubMed]

27. Miller, S.A.; Dykes, D.D.; Polesky, H.F. A simple salting out procedure for extracting DNA from human nucleated cells. *Nucleic Acids Res.* **1988**, *16*, 1215. [CrossRef]

28. Cawthon, R.M. Telomere measurement by quantitative PCR. *Nucleic Acids Res.* **2002**, *30*, e47. [CrossRef]

29. Scarabino, D.; Broggio, E.; Gambina, G.; Corbo, R.M. Leukocyte telomere length in mild cognitive impairment and Alzheimer's disease patients. *Exp. Gerontol.* **2017**, *98*, 143–147. [CrossRef]

30. Terwilliger, J.; Ott, J. *Handbook for Human Genetic Linkage*; John Hopkins University Press: Baltimore, MD, USA, 1994.

31. Wang, L.; Wei, Q.; Wang, L.-E.; Aldape, K.D.; Cao, Y.; Okcu, M.F.; Hess, K.R.; El-Zein, R.; Gilbert, M.R.; Woo, S.Y.; et al. Survival prediction in patients with glioblastoma multiforme by human telomerase genetic variation. *J. Clin. Oncol.* **2006**, *24*, 1627–1632. [CrossRef]

32. Jin, G.; Yoo, S.S.; Cho, S.; Jeon, H.-S.; Lee, W.-K.; Kang, H.-G.; Choi, Y.Y.; Choi, J.E.; Cha, S.-I.; Lee, E.B.; et al. Dual roles of a variable number of tandem repeat polymorphism in the *TERT* gene in lung cancer. *Cancer Sci.* **2011**, *102*, 144–149. [CrossRef] [PubMed]

33. Zhou, L.; Fu, G.; Wei, J.; Shi, J.; Pan, W.; Ren, Y.; Xiong, X.; Xia, J.; Shen, Y.; Li, H.; et al. The identification of two regulatory ESCC susceptibility genetic variants in the *TERT*-CLPTM1L loci. *Oncotarget* **2016**, *7*, 5495–5506. [CrossRef] [PubMed]

34. Martin-Ruiz, C.M.; Baird, D.; Roger, L.; Boukamp, P.; Krunic, D.; Cawthon, R.; Dokter, M.M.; van der Harst, P.; Bekaert, S.; de Meyer, T.; et al. Reproducibility of Telomere Length Assessment–An International Collaborative Study. *Int. J. Epidemiol.* **2015**, *44*, 1749–1754. [CrossRef] [PubMed]

35. Müezzinler, A.; Zaineddin, A.K.; Brenner, H. A systematic review of leukocyte telomere length and age in adults. *Ageing Res. Rev.* **2013**, *12*, 509–519. [CrossRef] [PubMed]

36. Mather, K.A.; Jorm, A.F.; Parslow, R.A.; Christensen, H. Is telomere length a biomarker of aging? A review. *J. Gerontol. A Biol. Sci. Med. Sci.* **2011**, *66*, 202–213. [CrossRef] [PubMed]

37. Mons, U.; Müezzinler, A.; Schöttker, B.; Dieffenbach, A.K.; Butterbach, K.; Schick, M.; Peasey, A.; De Vivo, I.; Trichopoulou, A.; Boffetta, P.; et al. Leukocyte Telomere Length and All-Cause, Cardiovascular Disease, and Cancer Mortality: Results from Individual-Participant-Data Meta-Analysis of 2 Large Prospective Cohort Studies. *Am. J. Epidemiol.* **2017**, *185*, 1317–1326. [CrossRef] [PubMed]

38. Wang, Q.; Zhan, Y.; Pedersen, N.L.; Fang, F.; Hägg, S. Telomere Length and All-Cause Mortality: A Meta-analysis. *Ageing Res. Rev.* **2018**, *48*, 11–20. [CrossRef] [PubMed]

39. Zhan, Y.; Liu, X.-R.; Reynolds, C.A.; Pedersen, N.L.; Hägg, S.; Clements, M.S. Leukocyte Telomere Length and All-Cause Mortality: A Between-Within Twin Study with Time-Dependent Effects Using Generalized Survival Models. *Am. J. Epidemiol.* **2018**, *187*, 2186–2191. [CrossRef]

40. Honig, L.S.; Kang, M.S.; Cheng, R.; Eckfeldt, J.H.; Thyagarajan, B.; Leiendecker-Foster, C.; Province, M.A.; Sanders, J.L.; Perls, T.; Christensen, K.; et al. Heritability of telomere length in a study of long-lived families. *Neurobiol. Aging* **2015**, *36*, 2785–2790. [CrossRef]

41. Bischoff, C.; Petersen, H.C.; Graakjaer, J.; Andersen-Ranberg, K.; Vaupel, J.W.; Bohr, V.A.; Kølvraa, S.; Christensen, K. No association between telomere length and survival among the elderly and oldest old. *Epidemiology* **2006**, *17*, 190–194. [CrossRef]

42. Martin-Ruiz, C.M.; Gussekloo, J.; van Heemst, D.; von Zglinicki, T.; Westendorp, R.G.J. Telomere length in white blood cells is not associated with morbidity or mortality in the oldest old: A population-based study. *Aging Cell* **2005**, *4*, 287–290. [CrossRef] [PubMed]

43. Svensson, J.; Karlsson, M.K.; Ljunggren, Ö.; Tivesten, Å.; Mellström, D.; Movérare-Skrtic, S. Leukocyte telomere length is not associated with mortality in older men. *Exp. Gerontol.* **2014**, *57*, 6–12. [CrossRef] [PubMed]

44. Franceschi, C.; Bonafè, M.; Valensin, S.; Olivieri, F.; De Luca, M.; Ottaviani, E.; De Benedictis, G. Inflammaging. An evolutionary perspective on immunosenescence. *Ann. N. Y. Acad. Sci.* **2000**, *908*, 244–254. [CrossRef] [PubMed]

45. Zhang, J.; Rane, G.; Dai, X.; Shanmugam, M.K.; Arfuso, F.; Samy, R.P.; Lai, M.K.P.; Kappei, D.; Kumar, A.P.; Sethi, G. Ageing and the telomere connection: An intimate relationship with inflammation. *Ageing Res. Rev.* **2016**, *25*, 55–69. [CrossRef] [PubMed]

GCAT TACG GCAT

genes

MDPI

Article

Inositol Polyphosphate Multikinase (*IPMK*), a Gene Coding for a Potential Moonlighting Protein, Contributes to Human Female Longevity

Francesco De Rango [1,†], **Paolina Crocco** [1,†], **Francesca Iannone** [1], **Adolfo Saiardi** [2], **Giuseppe Passarino** [1], **Serena Dato** [1,†] **and Giuseppina Rose** [1,†,*]

[1]   Department of Biology, Ecology and Earth Sciences, University of Calabria, 87036 Rende, Italy;
       francesco.derango@unical.it (F.D.R.); crocco.paola@gmail.com (P.C.); francescaiannonebio@gmail.com (F.I.);
       giuseppe.passarino@unical.it (G.P.); serena.dato@unical.it (S.D.)
[2]   MRC Laboratory for Molecular Cell Biology, University College London, London WC1E 6BT, UK;
       a.saiardi@ucl.ac.uk
*   Correspondence: pina.rose@unical.it; Tel.: +39-0984-492931
†   These authors have contributed equally to this work.

Received: 21 January 2019; Accepted: 4 February 2019; Published: 8 February 2019

Abstract: Biogerontological research highlighted a complex and dynamic connection between aging, health and longevity, partially determined by genetic factors. Multifunctional proteins with moonlighting features, by integrating different cellular activities in the space and time, may explain part of this complexity. Inositol Polyphosphate Multikinase (IPMK) is a potential moonlighting protein performing multiple unrelated functions. Initially identified as a key enzyme for inositol phosphates synthesis, small messengers regulating many aspects of cell physiology, IPMK is now implicated in a number of metabolic pathways affecting the aging process. IPMK regulates basic transcription, telomere homeostasis, nutrient-sensing, metabolism and oxidative stress. Here, we tested the hypothesis that the genetic variability of *IPMK* may affect human longevity. Single-SNP (single nuclear polymorphism), haplotype-based association tests as well as survival analysis pointed to the relevance of six out of fourteen genotyped SNPs for female longevity. In particular, haplotype analysis refined the association highlighting two SNPs, rs2790234 and rs6481383, as major contributing variants for longevity in women. Our work, the first to investigate the association between variants of *IPMK* and longevity, supports *IPMK* as a novel gender-specific genetic determinant of human longevity, playing a role in the complex network of genetic factors involved in human survival.

Keywords: aging; longevity; survival; SNP; polymorphism; IPMK; inositol phosphates; gender-specific association; moonlighting protein

1. Introduction

In the last few decades, research on aging has seen progressive growth due to the social and medical burden correlated to the increase of the elderly population in developed countries. These efforts point towards a better understanding of the connections between aging, health, and longevity as they may provide useful insights for strategies to improve the wellbeing of the elderly. The results obtained in different research areas underscored the dynamic complexity of such connections [1]. These studies often identify genes involved in the regulation of the aging process that are also susceptibility loci of one or multiple age-related diseases. For instance, many genetic variants associated with increased disease risk are present with high frequency among the oldest individuals: this means that a disease "risk allele" can also be a pro-longevity variant [2,3]. In some cases, the same variant exhibits opposite effects on the development of different diseases, with potential differential impact on longevity [4,5]. Finally, genetic risk factors may change their impact on mortality risk during the life course, i.e., from detrimental in middle life to beneficial at advanced ages, very often in a gender-specific way [6,7]. This makes the identification of genes that robustly associate with longevity very challenging; in fact, despite the high number of studies aimed at highlighting the genetic contributors to long-life, only *APOE* and *FOXO3A* were consistently replicated in different populations [8].

Among the mechanisms that may account for this complexity are interactions between different genes (epistasis) or single nuclear polymorphism (SNP)–SNP interactions at gene level, gene–environment (internal and external) interactions and pleiotropic (including trade-off-like) effects of genes on different phenotypes. Alongside this, potential contributing factors could be genes coding for proteins with different, relevant and often unrelated functions, since they may integrate various cellular activities in space and time. This special category of multifunctional proteins defined as moonlighting proteins [9], does not include protein isoforms resulting from different RNA splice variants, gene fusions or proteins with pleiotropic effects [10]. A protein with potential moonlighting capability is the inositol polyphosphate multikinase (IPMK).

This protein was initially discovered in budding yeast and named Arg82 for its ability to regulate arginine metabolism [11,12]. Mammalian IPMK has well-established roles in inositol phosphate metabolism as it converts inositol (1,4,5)-trisphosphate (IP$_3$) to IP$_4$ and IP$_4$ to IP$_5$ [13,14]. In addition to its kinase activity, IPMK can function as a nuclear phosphoinositide kinase (PI3-kinase), which produces PIP$_3$ from PIP$_2$ [15]. Through its PI3-kinase activity, IPMK activates Akt/PKB and its downstream signaling pathways [16]. In addition, it regulates several protein targets non-catalytically via protein–protein interactions, including cytosolic signaling factors such as the mammalian target of rapamycin complex 1 (mTORC1) [17] and the energy-sensing protein kinase AMPK [18]. Recently, Kim et al. [19] revealed that IPMK acts as an important regulator of Toll-like receptor (TLR)-induced innate immunity through its interaction with the tumor necrosis factor receptor–associated factor 6 (TRAF6). At the nuclear level, IPMK acts as a transcriptional coactivator for *p53* [20] and for serum response factor (SRF) signaling [21]. Finally, IPMK functions in the export of mRNA from the nucleus to the cytoplasm [22–24].

To our knowledge, there are no studies investigating the genetic variability of *IPMK* gene in human complex phenotypes; just one SNP, rs12570088, near to *IPMK* locus, was found related to the susceptibility to Alzheimer's and Crohn diseases [25].

Given the compelling evidence demonstrating IPMK's multifunctional nature and thus its classification as a moonlighting protein, the present paper addresses the hypothesis that *IPMK* genetic variability affects human aging and longevity.

2. Materials and Methods

2.1. Population Sample

We analysed five hundred sixty-eight unrelated subjects (252 men and 316 women aged 64–105 years), born in Calabria (South Italy) and recruited in the entire region through several campaigns, as previously reported [26]. Their Calabrian ancestry was ascertained up to the third generation. At baseline, all subjects were free of the major age-related pathologies (e.g., cancer, type-2 diabetes and cardiovascular diseases). The study was approved by the Ethical Committee of the University of Calabria (on 9-9-2004). Written informed consent was obtained from the subjects in accordance with institutional requirements and the Declaration of Helsinki principles.

The analyses were performed considering two sex- and age-specific groups obtained according to the survival functions of the Italian population from 1890 onward [27]. The two "thresholds of longevity" used to define these age classes were 88 years for men and 91 years for women. These cut-offs correspond to the point after which a significant negative change in the slope of the survival curve of the Italian population occurs. In particular, in the present study males younger than 88 and females younger than 91 years will be defined as controls (N = 309, mean age 74 years), while males older than 88 and females older than 91 years will be defined as cases (N = 259, mean age 96.9 years).

2.2. SNP Selection and Genotyping

We performed genotyping of 14 SNPs mapping within and nearby the *IPMK* gene, prioritized by a tagging approach. Analysis was performed by SEQUENOM MassArray iPLEX technology according to the procedure previously reported [28]. Sequenom Typer 4.0 Software was used for the management and analysis of the collected data. About 10% of the samples were reanalyzed and the concordance rate of the genotypes was higher than 99%.

2.3. Quality Control

After genotyping, samples were subjected to a battery of quality control (QC) tests. At sample level, subjects with a proportion of missing genotypes higher than 10% were dropped from the analysis. At SNP level, SNPs were excluded if they had a significant deviation from Hardy–Weinberg equilibrium (HWE, $p < 0.05$) in the control sample, a Missing Frequency (MiF) higher than 10% and a Minor Allele Frequency (MAF) lower than 5%.

2.4. Functional Parameters

2.4.1. Disability

A modification of the Katz' Index of activities of daily living (ADL) was used to assess the management of four everyday activities (toileting, getting up from bed, rising from a chair, walking around) [29]. For the analysis, ADL scores were dichotomized as 1 if the subject was able to perform every activity and 0 otherwise.

2.4.2. Physical Performance

Evaluation of Hand Grip strength (HG) was performed through a handheld dynamometer (SMEDLEY's dynamometer TTM, Tokyo, Japan) while the subject was sitting with the arm close to their body, by repeating the measure three times with the stronger hand. The maximum of these values was used in statistical analyses. When the test was not performed, it was indicated if it was because of physical disabilities or if the subject refused to participate.

2.4.3. Cognitive Functioning

Screening of cognitive impairment was carried out by MMSE, a 30-point scale able to evaluate several different cognitive areas including memory, calculation, abstraction, judgment, visual–spatial ability and language [30]. MMSE scores range from 0 (lowest cognitive function) to 30 (highest cognitive function). MMSE scores were normalized for age and educational status, variables known to affect the result of the test.

2.5. Statistical Methods

For each SNP, allele and genotype frequencies were estimated by gene counting from the observed genotypes. Hardy–Weinberg equilibrium (HWE) was tested by Fisher's exact test. Pairwise measures of linkage disequilibrium (LD) between the analyzed loci was estimated by Haploview (https://www.broadinstitute.org/haploview/haploview). A logistic regression model was also used to evaluate the effect of genetic variability on the chance to reach very advanced age. Different genetic models (dominant, additive and recessive) were used to test association, using for each SNP the minor allele as reference. For each SNP the most likely genetic model was then estimated on the basis of minimum level of statistical significance (Wald test *p*-value).

As this study was exploratory, the *p*-values are reported without employing conservative statistical significance thresholding procedure (e.g., Bonferroni correction) as that could eliminate potentially important findings.

In order to evaluate if the detected effect of the polymorphisms on longevity may result in differential patterns of survival of the different relevant genotypes, we evaluated survival after 10 years from the baseline visit. Univariate survival analysis was carried out by the Kaplan–Meier method and survival curves compared by log-rank test. Subjects alive after the follow-up time were considered as censored, and this time was used as the censoring date in the survival analyses. In addition, hazard ratios (HR) and 95% confidence intervals (95% CI) were estimated by using Cox proportional hazard models taking into account age as a confounder variable.

Pairwise measures of linkage disequilibrium (LD) between the analyzed loci were calculated by Plink 1.9 [31] and plotted with the Haploview version 4.2 [32]. The amount of LD was quantified by Lewontin's coefficient (D'). Haplotype-based association analysis within the generalized linear model (GLM) framework was used to model the effect of haplotypes on the probability to attain longevity, by the haplo.stats package of R. The haplo.score function of this package has been used to obtain the score statistics. Permutation-based *p*-values were used to evaluate the significance of the scores obtained (10,000 permutations).

Statistical analyses have been performed using SNPassoc and surv packages of R [33].

3. Results

Fourteen SNPs from about 76 kb genomic sequences spanning the *IPMK* gene were selected for examination by a tagging approach. The QC phase excluded three SNPs. In particular, two SNP were excluded due to MiF data higher than 10% (rs1698392, rs2440854) and one because it did not satisfy the HWE (rs2275443). Figure 1A shows the eleven high quality SNPs that were tested for association with longevity with their corresponding gene position, while panel B depicts the degree of LD between pairs of SNPs.

3.1. Association with Longevity

3.1.1. Single SNP Analysis

The general characteristics of the analyzed sample are described in Table 1. Since a number of studies highlighted gender- and age-specific associations with survival at advanced age, in this work we analyzed the role of *IPMK* SNPs in the predisposition to become long-lived in gender subgroups.

Table 1. General characteristics and post-survey mortality in the analyzed sample.

	Elderly Subjects	Long-Lived Subjects
N (age)	309 (74.06 ± 6.95)	259 (96.92 ± 3.72)
Females %	49.5%	63.0%
Height (cm)	160.6 (9.7)	151.4 (9.5)
BMI	26.9 (4.2)	23.21 (4.1)
HG strength [Kg (SD)]	21.89 (9.9)	13.04 (6.4)
ADL* (% Disabled)	17%	69%
MMSE	23.3 (5.4)	14.0 (6.8)

ADL, Activity Daily Living; HG, Hand Grip; BMI, Body Mass Index; MMSE, Mini Mental State Examination; for each parameter, mean value and standard deviation, in brackets, are shown. *Participants were defined as "not disabled" if independent in all items and "disabled" if dependent in at least one item.

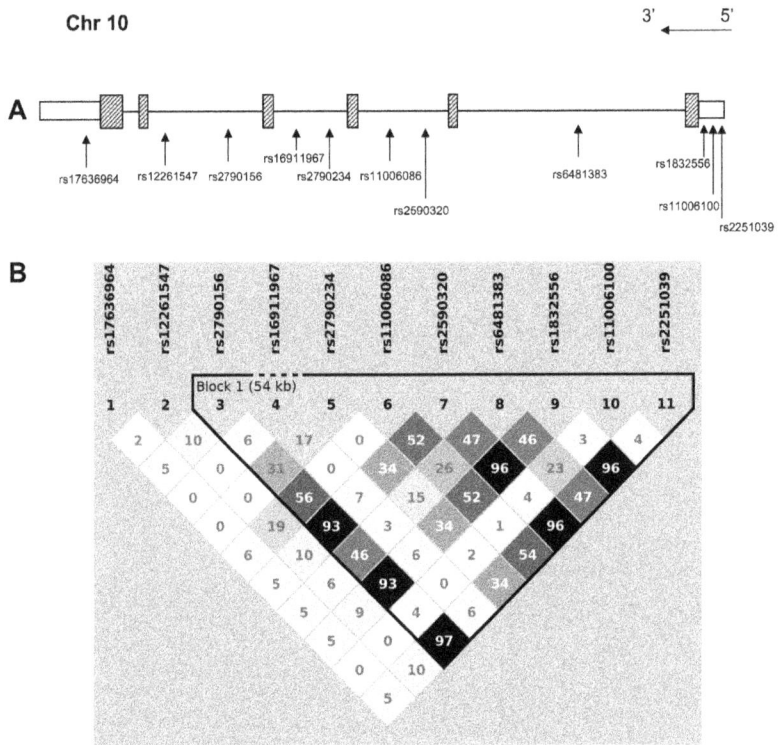

Figure 1. Schematic representation of (**A**) selected polymorphisms in the Inositol Polyphosphate Multikinase (*IPMK*) region; (**B**) linkage disequilibrium (r^2 coefficient) among the single nuclear polymorphisms (SNPs).

From the results of the logistic regression analysis shown in Table 2, it can be seen that significant differences between the long-lived subjects and younger controls are present among females only. In particular, six out of eleven markers (in order: rs2790156-G/A, rs2790234-C/G, rs2590320-C/A, rs6481383-C/T, rs1832556-G/A, rs2251039-C/T) were significantly associated with the longevity phenotype under a dominant model of inheritance. For all the SNPs, the presence of the minor allele conferred decreased odds to reach advanced old age. rs2790234 showed the greatest impact (Odd Ratio, OR, 0.33; 95% Confidence Interval, CI, 0.16–0.67; P = 0.00225), while association of similar magnitude was observed for the other five polymorphisms with ORs (95% CI) of 0.62, 0.572, 0.592, 0.592 and 0.612 (all *p*-values < 0.05) for rs2790156, rs2590320, rs6481383, rs1832556, rs2251039, respectively).

3.1.2. Haplotype-Based Analysis

To further explore the association of the entire region with longevity, we performed a haplotype analysis among the six SNPs associated with longevity. As shown in Figure 1B, all six SNPs lie in a large LD block, with rs2790156, rs2590320, rs1832556 and rs2251039 in strong LD and rs2790234 and rs6481383 in a weak linkage. Among all possible haplotypes, we found only four combinations: G-C-C-C-G-C, 61%; G-C-C-T-G-C, 16%; A-C-A-T-A-T, 14%; A-G-A-T-A-T, 7% (Table 3). In line with the single locus analysis, we found a negative association of the minor allele combination A-G-A-T-A-T with longevity in females (*p*-value = 0.002). On the contrary, a positive association was observed for the opposite combination G-C-C-C-G-C (*p*-value = 0.024). A deep analysis of the associated haplotypes showed that the strength of these associations was influenced by the allelic status at rs2790234 and rs6481383. Indeed, while A-**G**-A-T-A-T is significantly associated, the A-**C**-A-T-A-T is not; likewise, while G-C-C-**C**-G-C showed an effect on longevity, this was not true for G-C-C-**T**-G-C.

Table 2. Results of the logistic regression models for *IPMK* SNPs in the sample divided by sex.

(a) Females			
SNP (Major/Minor Allele)	OR	95% CI	*p*-value
rs17636964 (G/C)	1.48	0.82–2.662	0.185
rs12261547 (G/C)	1.39	0.62–3.11	0.415
rs2790156 (G/A)	0.61	0.38–0.98	0.042
rs16911967 (G/C)	0.40	0.12–1.33	0.136
rs2790234 (C/G)	0.33	0.16–0.67	0.002
rs11006086 (T/C)	0.67	0.34–1.32	0.255
rs2590320 (C/A)	0.57	0.36–0.91	0.019
rs6481383 (C/T)	0.59	0.37–0.94	0.026
rs1832556 (G/A)	0.59	0.37–0.94	0.028
rs11006100 (T/A)	0.69	0.42–1.14	0.154
rs2251039 (C/T)	0.61	0.38–0.97	0.038
(b) Males			
SNP	OR	95% CI	*p*-value
rs17636964 (G/C)	0.98	0.50–1.94	0.974
rs12261547 (G/C)	0.52	0.16–1.66	0.272
rs2790156 (G/A)	0.79	0.46–1.35	0.397
rs16911967 (G/C)	2.08	0.54–7.96	0.283
rs2790234 (C/G)	0.98	0.49–1.94	0.959
rs11006086 (T/C)	0.76	0.31–1.85	0.550
rs2590320 (C/A)	0.81	0.48–1.37	0.436
rs6481383 (C/T)	1.10	0.65–1.87	0.713
rs1832556 (G/A)	0.81	0.47–1.37	0.436
rs11006100 (T/A)	1.73	0.98–3.04	0.056
rs2251039 (C/T)	0.78	0.46–1.34	0.377

OR: Odd Ratio; CI: Confidence Interval.

Table 3. Estimation of haplotype frequencies in the IPMK SNPs (in order: rs2790156, rs2790234, rs2590320, rs6481383, rs1832556, rs2251039) and association with longevity in the female sample.

Haplotype	Frequency	Score	*p*-Value *
A-G-A-T-A-T	0.067	−2.897	0.002
A-C-A-T-A-T	0.138	−0.668	0.483
G-C-C-T-G-C	0.161	−0.353	0.715
G-C-C-C-G-C	0.616	2.155	0.024

* simulated *p*-value obtained by Monte Carlo replication up to 10,000 bootstraps.

3.2. Association with Survival

By using 10 years of follow-up survival data, we investigated if the single variants associated with female longevity also influenced the survival of the younger cohort. As shown in Figure 2, consistent with the detrimental effect on longevity, we found a trend toward significance for three out of six variants associated with longevity, rs2590320, rs1832556 and rs2251039, with HR values 1.72 (0.91–3.23), 1.75 (0.93–3.28), 1.75 (0.93–3.28) respectively ($p < 0.1$). we could not perform a haplotype-based survival analysis because a classification of carriers or non- carriers would reduce the size of the two classes too much.

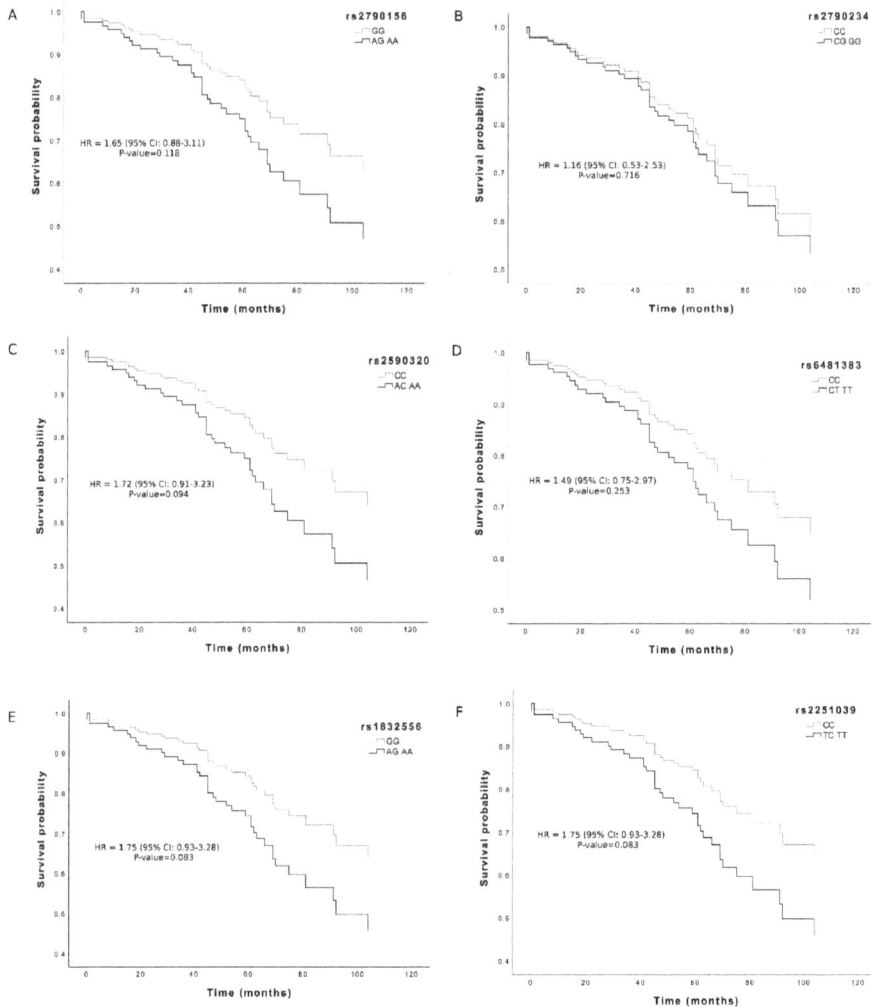

Figure 2. Survival functions of female carriers of minor allele (black) vs non-carriers (grey) of IPMK variants. (**A**) rs2790156; (**B**) rs2790234; (**C**) rs2590320; (**D**) rs6481383; (**E**) rs1832556; (**F**) rs2251039. Time is expressed in months, where 0 is considered the time of recruitment and each individual is followed up for survival status till death. The Cox regression was adjusted for age. Hazard ratio (HR) value, confidence interval and *p*-value from Cox regression analysis are reported inside the figure.

3.3. Association with Functional Parameters

To investigate whether the variants in *IPMK* gene also concur to determine the age-related physiological decline, we analyzed the SNPs in relation to markers of physical (ADL and Hand Grip) and cognitive (MMSE) performance. No significant association was detected (data not shown). As a result, we conclude that *IPMK* has an effect on survival and longevity independently of the tested variables.

4. Discussion

Our study was designed to test the hypothesis that genetic variability at the *IPMK* locus contributes to survival to very old age. We provided evidence that polymorphisms in this gene significantly affect the females' chance of survival to old age, a result that implicates IPMK, a multifunctional protein with potential moonlighting functions, as a significant contributor to gender differences in longevity.

The gender difference in life expectancy and mortality, including survival to extreme age, as well as prevalence and incidence of the most important age-related diseases, is supported by a huge amount of clinical and demographic data [1,34]. In almost all modern populations, females live longer than males and this has been attributed to a particular combination of genetic factors, environmental factors (nutrition and stress), sex hormones and immunity, along with socio-economic and cultural factors [35,36]. Gender-specific longevity alleles were identified for a long time [37,38] and recently confirmed by genome-wide association studies (GWAs) [39,40]. These studies also indicated that different pathways contribute to longevity in men and women; for instance, paths involved in inflammation and immunity emerged as male-specific, while those involved in PGC-1α (PPARγ coactivator-1α) function and tryptophan metabolism emerged as female specific [40]. It is intriguing that many members of these pathways are known to perform diverse unrelated functions, behaving as moonlighting proteins [41,42]. As previously discussed, proteins with moonlighting properties may, in part, explain the complex role of genetic factors in determining the longevity phenotype, including gender-linked effects. In this sense, IPMK is surely a possible player because of multifunctional protein feature. IPMK catalyzes key steps leading to the synthesis of inositol pyrophosphates [43]. These molecules, that as the name indicates contain one or more pyrophosphate moiety, control several aspects of cell physiology essential for cell survival. Inositol pyrophosphate regulate telomere length [44], vesicular trafficking [45], DNA recombination [46], ROS signaling [47] and energetic metabolism [48], likely by controlling cellular phosphate homeostasis [49,50]. In fact, these molecules regulate the pathophysiology of metabolic disorders such diabetes and obesity [51,52]. The numerous and distinct connections between metabolism and aging that research is highlighting [53,54] suggest that inositol pyrophosphate might be relevant to the aging process. Remarkably, the knockout of inositol hexakisphosphate kinase 3 (IP6K3), an essential enzyme for inositol pyrophosphates synthesis, results in altered metabolism and extends mice lifespan [55]. Furthermore, we associated two SNPs in the 5'-flanking promoter region of the *IP6K3* gene with the susceptibility to late onset Alzheimer's disease (LOAD) [28]. Therefore, the catalytic activity of IPMK, indispensable for inositol pyrophosphate synthesis, could be ultimately important for controlling metabolism and lifespan. Independently from its enzymatic activity, through protein–protein interaction, IPMK acts as a signaling hub in regulating nutrient and energetic pathways, including mTOR and AMPK [17,56]. The inhibition of mTORC1 pathway components extends lifespan and confers protection against an increasing list of age-related diseases, while on the contrary its over-induction leads to a higher risk of age-related diseases and decreased lifespan [57,58] and references therein]. A role in the control of cell survival and death has been also highlighted. A small deletion that leads a truncated form of IMPK was found to reduce the activation of p53 and increase the resistance to apoptosis of cancer cell lines [59]. Moreover, Davey et al. found that IPMK plays an important role in necroptosis, a form of regulated cell death prompted by injury and infection [60]. Both apoptosis and necroptosis impact on a variety of processes governing cell physiology and homeostasis with implications in health and disease [61].

It is interesting to note that many of the processes in which IPMK participates influence aging in a gender specific manner. For instance, some authors have suggested that gender differences in lifespan are due to gender-specific susceptibility to oxidative stress [62] and yet gender-specific survival is associated with sex differences in telomere dynamics [63]. It has also been demonstrated that rapamycin, the inhibitor of mTOR signaling, extends lifespan in dose- and gender-specific ways [64] and that there are sex-differences in muscle AMPK activation [65].

On the whole, these studies corroborate our findings that six *IPMK* SNPs significantly affect female longevity. Haplotype analysis confirmed the single SNP analyses, identifying an advantageous effect in carriers of the haplotype G-C-C-C-G-C; conversely the A-G-A-T-A-T haplotype is a disadvantageous combination for longevity. Haplotypes analysis allowed us to establish that among the six SNPs, two SNPs, rs2790234-C/G and rs6481383-C/T, were more likely to have an effect on lifespan, the minor rs2790234-G allele conferring a longevity disadvantage and the major rs6481383-C allele having a beneficial effect on the trait. The absence of correlation with survival and parameters of quality of aging suggests that these SNPs play a major role on the probability to achieve longevity.

Since we have no demonstrable functional explanation for the association as both SNPs occur in large intronic regions, it is difficult to evaluate their significance at this time. Moreover, both SNPs lie in a large region of LD, which likely contain hundreds of polymorphisms with one or more others that may have a direct contribution. Thus, our findings could be relevant for future investigations.

This study has some limitations that merit consideration. First, the sample size is rather small, which limits its statistical power. The sample size might have influenced the significance of survival analysis, so only trends have been identified. However, this result may also depend on the follow-up time of 10 years, not sufficient to draw long-term conclusions on the effect of genetic variants with a minor effect on survival. Therefore, the sample size should be increased and further explorations in additional populations and other countries are needed before drawing further conclusions. Another possible drawback is the lack of a proper correction for multiple testing. However, it should be noted that this was a pilot study, the first to analyze the association of *IPMK* variability in human aging and longevity, so a Bonferroni correction would have eliminated potentially important findings if applied. Furthermore, because associated SNPs are intronic, experimental evidence in support of the hypothesis that the detrimental genotypes influence the function of the protein should be carried out.

5. Conclusions

Specific *IPMK* haplotypes affect lifespan in women. Although our studies do not allow definitive conclusions, we believe that our findings can provide a basis for future studies to better clarify the basic mechanisms linking *IPMK* to female longevity and potential targets for realizing gender-specific therapeutic interventions. Finally, we believe that proteins with moonlighting capabilities, such as IPMK, could represent one of the factors that makes it difficult to disentangle the genetic complexity of the longevity phenotype.

Author Contributions: Conceptualization, G.R. and A.S.; methodology, P.C. and F.I.; formal analysis, F.D.R.; data curation, F.D.R.; writing—original draft preparation, G.R. and S.D.; writing—review and editing, G.R., G.P., S.D., P.C. and A.S.; funding acquisition, G.R.

Funding: This research was funded by the Italian Ministry of University and Research (PRIN: Progetti di Ricerca di rilevante Interesse Nazionale – 2015, Prot. 20157ATSLF) to G.R.

Acknowledgments: The work has been made possible by the collaboration with Gruppo Baffa (Sadel Spa, Sadel San Teodoro srl, Sadel CS srl, Casa di Cura Madonna dello Scoglio, AGI srl, Casa di Cura Villa del Rosario srl, Savelli Hospital srl, Casa di Cura Villa Ermelinda).

Conflicts of Interest: The authors declare no conflict of interest.

References

1. Ukraintseva, S.; Yashin, A.; Arbeev, K.; Kulminski, A.; Akushevich, I.; Wu, D.; Joshi, G.; Land, K.C.; Stallard, E. Puzzling role of genetic risk factors in human longevity: "Risk alleles" as pro-longevity variants. *Biogerontology* **2016**, *17*, 109–127. [CrossRef]
2. Mooijaart, S.P.; van Heemst, D.; Noordam, R.; Rozing, M.P.; Wijsman, C.A.; de Craen, A.J.; Westendorp, R.G.; Beekman, M.; Slagboom, P.E. Polymorphisms associated with type 2 diabetes in familial longevity: The Leiden Longevity Study. *Aging* **2011**, *3*, 55–62. [CrossRef] [PubMed]

3. Sebastiani, P.; Riva, A.; Montano, M.; Pham, P.; Torkamani, A.; Scherba, E.; Benson, G.; Milton, J.N.; Baldwin, C.T.; Andersen, S.; et al. Whole genome sequences of a male and female supercentenarian, ages greater than 114 years. *Front. Genet.* **2011**, *2*, 90. [CrossRef]

4. Lio, D.; Scola, L.; Crivello, A.; Colonna-Romano, G.; Candore, G.; Bonafè, M.; Cavallone, L.; Franceschi, C.; Caruso, C. Gender-specific association between -1082 IL-10 promoter polymorphism and longevity. *Genes Immun.* **2002**, *3*, 30–33. [CrossRef]

5. Kulminski, A.M.; Culminskaya, I.; Arbeev, K.G.; Ukraintseva, S.V.; Arbeeva, L.; Yashin, A.I. Trade-off in the effect of the APOE gene on the ages at onset of cardiovascular disease and cancer across ages, gender, and human generations. *Rejuvenation Res.* **2013**, *16*, 28–34. [CrossRef] [PubMed]

6. Petersen, L.K.; Christensen, K.; Kragstrup, J. Lipid-lowering treatment to the end? A review of observational studies and RCTs on cholesterol and mortality in 80+-year olds. *Age Ageing* **2010**, *39*, 674–680. [CrossRef] [PubMed]

7. van Vliet, P.; Oleksik, A.M.; van Heemst, D.; de Craen, A.J.M.; Westendorp, R.G.J. Dynamics of traditional metabolic risk factors associate with specific causes of death in old age. *J. Gerontol. Ser. A* **2010**, *65*, 488–494. [CrossRef] [PubMed]

8. Zeng, Y.; Nie, C.; Min, J.; Liu, X.; Li, M.; Chen, H.; Xu, H.; Wang, M.; Ni, T.; Li, Y.; et al. Novel loci and pathways significantly associated with longevity. *Sci. Rep.* **2016**, *6*, 21243. [CrossRef] [PubMed]

9. Jeffery, C.J. Moonlighting proteins. *Trends Biochem. Sci.* **1999**, *24*, 8–11. [CrossRef]

10. Jeffery, C.J. Molecular mechanisms for multitasking: recent crystal structures of moonlighting proteins. *Curr. Opin. Struct. Biol.* **2004**, *14*, 663–668. [CrossRef]

11. Bosch, D.; Saiardi, A. Arginine transcriptional response does not require inositol phosphate synthesis. *J. Biol. Chem.* **2012**, *287*, 38347–38355. [CrossRef] [PubMed]

12. Hatch, A.J.; Odom, A.R.; York, J.D. Inositol phosphate multikinase dependent transcriptional control. *Adv. Biol. Regul.* **2017**, *64*, 9–19. [CrossRef] [PubMed]

13. Saiardi, A.; Nagata, E.; Luo, H.R.; Sawa, A.; Luo, X.; Snowman, A.M.; Snyder, S.H. Mammalian inositol polyphosphate multikinase synthesizes inositol 1,4,5-trisphosphate and an inositol pyrophosphate. *Proc. Natl. Acad. Sci. USA* **2001**, *98*, 2306–2311. [CrossRef] [PubMed]

14. Seacrist, C.D.; Blind, R.D. Crystallographic and kinetic analyses of human IPMK reveal disordered domains modulate ATP binding and kinase activity. *Sci. Rep.* **2018**, *8*, 16672. [CrossRef] [PubMed]

15. Resnick, A.C.; Snowman, A.M.; Kang, B.N.; Hurt, K.J.; Snyder, S.H.; Saiardi, A. Inositol polyphosphate multikinase is a nuclear PI3-kinase with transcriptional regulatory activity. *Proc. Natl. Acad. Sci. USA* **2005**, *102*, 12783–12788. [CrossRef] [PubMed]

16. Maag, D.; Maxwell, M.J.; Hardesty, D.A.; Boucher, K.L.; Choudhari, N.; Hanno, A.G.; Ma, J.F.; Snowman, A.S.; Pietropaoli, J.W.; Xu, R.; et al. Inositol polyphosphate multikinase is a physiologic PI3-kinase that activates Akt/PKB. *Proc. Natl. Acad. Sci. USA* **2011**, *108*, 1391–1396. [CrossRef] [PubMed]

17. Kim, S.; Kim, S.F.; Maag, D.; Maxwell, M.J.; Resnick, A.C.; Juluri, K.R.; Chakraborty, A.; Koldobskiy, M.A.; Cha, S.H.; Barrow, R.; et al. Amino acid signaling to mTOR mediated by inositol polyphosphate multikinase. *Cell Metab.* **2011**, *2*, 215–221. [CrossRef]

18. Dailey, M.J.; Kim, S. Inositol polyphosphate multikinase: An emerging player for the central action of AMP-activated protein kinase. *Biochem. Biophys. Res. Commun.* **2012**, *421*, 1–3. [CrossRef]

19. Kim, E.; Beon, J.; Lee, S.; Park, S.J.; Ahn, H.; Kim, M.G.; Park, J.E.; Kim, W.; Yuk, J.M.; Kang, S.J.; et al. Inositol polyphosphate multikinase promotes Toll-like receptor-induced inflammation by stabilizing TRAF6. *Sci. Adv.* **2017**, *3*, e1602296. [CrossRef]

20. Xu, R.; Sen, N.; Paul, B.D.; Snowman, A.M.; Rao, F.; Vandiver, M.S.; Xu, J.; Snyder, S.H. Inositol polyphosphate multikinase is a coactivator of p53-mediated transcription and cell death. *Sci. Signal.* **2013**, *6*, ra22. [CrossRef]

21. Kim, E.; Tyagi, R.; Lee, J.Y.; Park, J.; Kim, Y.R.; Beon, J.; Chen, P.Y.; Cha, J.Y.; Snyder, S.H.; Kim, S. Inositol polyphosphate multikinase is a coactivator for serum response factor-dependent induction of immediate early genes. *Proc. Natl. Acad. Sci. USA* **2013**, *110*, 19938–19943. [CrossRef] [PubMed]

22. Saiardi, A.; Caffrey, J.J.; Snyder, S.H.; Shears, S.B. Inositol polyphosphate multikinase (ArgRIII) determines nuclear mRNA export in Saccharomyces cerevisiae. *FEBS Lett.* **2000**, *468*, 28–32. [CrossRef]

23. Carmody, S.R.; Wente, S.R. mRNA nuclear export at a glance. *J. Cell. Sci.* **2009**, *122*, 1933–1937. [CrossRef] [PubMed]

24. Kim, E.; Beon, J.; Lee, S.; Park, J.; Kim, S. IPMK: A versatile regulator of nuclear signaling events. *Adv. Biol. Regul.* **2016**, *61*, 25–32. [CrossRef] [PubMed]
25. Yokoyama, J.S.; Wang, Y.; Schork, A.J.; Thompson, W.K.; Karch, C.M.; Cruchaga, C.; McEvoy, L.K.; Witoelar, A.; Chen, C.H.; Holland, D.; et al. Alzheimer's Disease Neuroimaging Initiative. Association between Genetic Traits for Immune-Mediated Diseases and Alzheimer Disease. *JAMA Neurol.* **2016**, *73*, 691–697. [CrossRef] [PubMed]
26. De Rango, F.; Montesanto, A.; Berardelli, M.; Mazzei, B.; Mari, V.; Lattanzio, F.; Corsonello, A.; Passarino, G. To grow old in southern Italy: a comprehensive description of the old and oldest old in Calabria. *Gerontology* **2011**, *57*, 327–334. [CrossRef] [PubMed]
27. Passarino, G.; Montesanto, A.; De Rango, F.; Garasto, S.; Berardelli, M.; Domma, F.; Mari, V.; Feraco, E.; Franceschi, C.; De Benedictis, G. A cluster analysis to define human aging phenotypes. *Biogerontology* **2007**, *8*, 283–290. [CrossRef]
28. Crocco, P.; Saiardi, A.; Wilson, M.S.; Maletta, R.; Bruni, A.C.; Passarino, G.; Rose, G. Contribution of polymorphic variation of inositol hexakisphosphate kinase 3 (IP6K3) gene promoter to the susceptibility to late onset Alzheimer's disease. *Biochim. Biophys. Acta.* **2016**, *1862*, 1766–1773. [CrossRef]
29. Katz, S.; Ford, A.B.; Moskowitz, R.W.; Jackson, B.A.; Jaffe, M.W. Studies of illness in the aged. The index of Adl: A standardized measure of biological and psychosocial function. *JAMA* **1963**, *185*, 914–919. [CrossRef]
30. Folstein, M.F.; Folstein, S.E.; McHugh, P.R. "Mini-mental state". A practical method for grading the cognitive state of patients for the clinician. *J. Psychiatr. Res.* **1975**, *12*, 189–198. [CrossRef]
31. Plink 1.9 home. Available online: https://www.cog-genomics.org/plink2 (accessed on 8 February 2019).
32. Barrett, J.C.; Fry, B.; Maller, J.; Daly, M.J. Haploview: Analysis and visualization of LD and haplotype maps. *Bioinformatics* **2005**, *21*, 263–265. [CrossRef] [PubMed]
33. The R Project for Statistical Computing. Available online: http://www.R-project.org/ (accessed on 8 February 2019).
34. Austad, S.N.; Fischer, K.E. Sex Differences in Lifespan. *Cell Metab.* **2016**, *23*, 1022–1033. [CrossRef] [PubMed]
35. Ostan, R.; Monti, D.; Gueresi, P.; Bussolotto, M.; Franceschi, C.; Baggio, G. Gender, aging and longevity in humans: An update of an intriguing/neglected scenario paving the way to a gender-specific medicine. *Clin. Sci. (Lond)* **2016**, *130*, 1711–1725. [CrossRef] [PubMed]
36. Dato, S.; Rose, G.; Crocco, P.; Monti, D.; Garagnani, P.; Franceschi, C.; Passarino, G. The genetics of human longevity: An intricacy of genes, environment, culture and microbiome. *Mech. Ageing Dev.* **2017**, *165*, 147–155. [CrossRef] [PubMed]
37. Garasto, S.; Rose, G.; Derango, F.; Berardelli, M.; Corsonello, A.; Feraco, E.; Mari, V.; Maletta, R.; Bruni, A.; Franceschi, C.; et al. The study of APOA1, APOC3 and APOA4 variability in healthy ageing people reveals another paradox in the oldest old subjects. *Ann. Hum. Genet.* **2003**, *67*, 54–62. [CrossRef] [PubMed]
38. Altomare, K.; Greco, V.; Bellizzi, D.; Berardelli, M.; Dato, S.; De Rango, F.; Garasto, S.; Rose, G.; Feraco, E.; Mari, V.; et al. The allele (A) (-110) in the promoter region of the HSP70-1 gene is unfavorable to longevity in women. *Biogerontology* **2003**, *4*, 215–220. [CrossRef] [PubMed]
39. Deelen, J.; Beekman, M.; Uh, H.W.; Helmer, Q.; Kuningas, M.; Christiansen, L.; Kremer, D.; van der Breggen, R.; Suchiman, H.E.; Lakenberg, N.; et al. Genome-wide association study identifies a single major locus contributing to survival into old age; the APOE locus revisited. *Aging Cell* **2011**, *10*, 686–698. [CrossRef]
40. Zeng, Y.; Nie, C.; Min, J.; Chen, H.; Liu, X.; Ye, R.; Chen, Z.; Bai, C.; Xie, E.; Yin, Z.; et al. Sex Differences in Genetic Associations With Longevity. *JAMA Netw. Open* **2018**, *1*, e181670. [CrossRef]
41. Jeffery, C.J. Moonlighting proteins–an update. *Mol Biosyst.* **2009**, *5*, 345–350. [CrossRef]
42. Houtkooper, R.H.; Williams, R.W.; Auwerx, J. Metabolic networks of longevity. *Cell* **2010**, *142*, 9–14. [CrossRef]
43. Resnick, A.C.; Saiardi, A. Inositol polyphosphate multikinase: Metabolic architect of nuclear inositides. *Front. Biosci.* **2008**, *13*, 856–866. [CrossRef] [PubMed]
44. Saiardi, A.; Resnick, A.C.; Snowman, A.M.; Wendland, B.; Snyder, S.H. Inositol pyrophosphates regulate cell death and telomere length through phosphoinositide 3-kinase-related protein kinases. *Proc. Natl. Acad. Sci. USA* **2005**, *102*, 1911–1914. [CrossRef] [PubMed]
45. Saiardi, A.; Sciambi, C.; McCaffery, J.M.; Wendland, B.; Snyder, S.H. Inositol pyrophosphates regulate endocytic trafficking. *Proc. Natl. Acad. Sci. USA* **2002**, *99*, 14206–14211. [CrossRef] [PubMed]

46. Jadav, R.S.; Chanduri, M.V.; Sengupta, S.; Bhandari, R. Inositol pyrophosphate synthesis by inositol hexakisphosphate kinase 1 is required for homologous recombination repair. *J. Biol Chem.* **2013**, *288*, 3312–3321. [CrossRef] [PubMed]

47. Onnebo, S.M.; Saiardi, A. Inositol pyrophosphates modulate hydrogen peroxide signalling. *Biochem. J.* **2009**, *423*, 109–118. [CrossRef] [PubMed]

48. Szijgyarto, Z.; Garedew, A.; Azevedo, C.; Saiardi, A. Influence of inositol pyrophosphates on cellular energy dynamics. *Science* **2011**, *334*, 802–805. [CrossRef] [PubMed]

49. Wild, R.; Gerasimaite, R.; Jung, J.Y.; Truffault, V.; Pavlovic, I.; Schmidt, A.; Saiardi, A.; Jessen, H.J.; Poirier, Y.; Hothorn, M.; et al. Control of eukaryotic phosphate homeostasis by inositol polyphosphate sensor domains. *Science* **2016**, *352*, 986–990. [CrossRef]

50. Azevedo, C.; Saiardi, A. Eukaryotic Phosphate Homeostasis: The Inositol Pyrophosphate Perspective. *Trends Biochem. Sci.* **2017**, *42*, 219–231. [CrossRef]

51. Illies, C.; Gromada, J.; Fiume, R.; Leibiger, B.; Yu, J.; Juhl, K.; Yang, S.N.; Barma, D.K.; Falck, J.R.; Saiardi, A.; et al. Requirement of inositol pyrophosphates for full exocytotic capacity in pancreatic beta cells. *Science* **2007**, *318*, 1299–1302. [CrossRef]

52. Zhu, Q.; Ghoshal, S.; Rodrigues, A.; Gao, S.; Asterian, A.; Kamenecka, T.M.; Barrow, J.C.; Chakraborty, A. Adipocyte-specific deletion of Ip6k1 reduces diet-induced obesity by enhancing AMPK-mediated thermogenesis. *J. Clin. Invest.* **2016**, *126*, 4273–4288. [CrossRef]

53. Finkel, T. The metabolic regulation of aging. *Nat. Med.* **2015**, *21*, 1416–1423. [CrossRef]

54. Dato, S.; Bellizzi, D.; Rose, G.; Passarino, G. The impact of nutrients on the aging rate: A complex interaction of demographic; environmental and genetic factors. *Mech. Ageing Dev.* **2016**, *154*, 49–61. [CrossRef] [PubMed]

55. Moritoh, Y.; Oka, M.; Yasuhara, Y.; Hozumi, H.; Iwachidow, K.; Fuse, H.; Tozawa, R. Inositol Hexakisphosphate Kinase 3 Regulates Metabolism and Lifespan in Mice. *Sci. Rep.* **2016**, *6*, 32072. [CrossRef] [PubMed]

56. Bang, S.; Chen, Y.; Ahima, R.S.; Kim, S.F. Convergence of IPMK and LKB1-AMPK signaling pathways on metformin action. *Mol. Endocrinol.* **2014**, *28*, 1186–1193. [CrossRef] [PubMed]

57. Stanfel, M.N.; Shamieh, L.S.; Kaeberlein, M.; Kennedy, B.K. The TOR pathway comes of age. *Biochim. Biophys. Acta* **2009**, *1790*, 1067–1074. [CrossRef] [PubMed]

58. Dato, S.; Hoxha, E.; Crocco, P.; Iannone, F.; Passarino, G.; Rose, G. Amino acids and amino acid sensing: Implication for aging and diseases. *Biogerontology* **2019**, *20*, 17–31. [CrossRef] [PubMed]

59. Sei, Y.; Zhao, X.; Forbes, J.; Szymczak, S.; Li, Q.; Trivedi, A.; Voellinger, M.; Joy, G.; Feng, J.; Whatley, M.; et al. A Hereditary Form of Small Intestinal Carcinoid Associated with a Germline Mutation in Inositol Polyphosphate Multikinase. *Gastroenterology* **2015**, *149*, 67–78. [CrossRef]

60. Dovey, C.M.; Diep, J.; Clarke, B.P.; Hale, A.T.; McNamara, D.E.; Guo, H.; Brown, N.W., Jr.; Cao, J.Y.; Grace, C.R.; Gough, P.J.; et al. MLKL Requires the Inositol Phosphate Code to Execute Necroptosis. *Mol. Cell.* **2018**, *70*, 936–948.e7. [CrossRef]

61. Tower, J. Programmed cell death in aging. *Ageing Res. Rev.* **2015**, *23*, 90–100. [CrossRef]

62. Kander, M.C.; Cui, Y.; Liu, Z. Gender difference in oxidative stress: A new look at the mechanisms for cardiovascular diseases. *J. Cell Mol. Med.* **2017**, *21*, 1024–1032. [CrossRef]

63. Gardner, M.; Bann, D.; Wiley, L.; Cooper, R.; Hardy, R.; Nitsch, D.; Martin-Ruiz, C.; Shiels, P.; Sayer, A.A.; Barbieri, M.; et al. Halcyon study team. Gender and telomere length: Systematic review and meta-analysis. *Exp. Gerontol.* **2014**, *51*, 15–27. [CrossRef] [PubMed]

64. Miller, R.A.; Harrison, D.E.; Astle, C.M.; Fernandez, E.; Flurkey, K.; Han, M.; Javors, M.A.; Li, X.; Nadon, N.L.; Nelson, J.F.; et al. Rapamycin-mediated lifespan increase in mice is dose and sex dependent and metabolically distinct from dietary restriction. *Aging Cell* **2014**, *13*, 468–477. [CrossRef] [PubMed]

65. Roepstorff, C.; Thiele, M.; Hillig, T.; Pilegaard, H.; Richter, E.A.; Wojtaszewski, J.F.; Kiens, B. Higher skeletal muscle alpha2AMPK activation and lower energy charge and fat oxidation in men than in women during submaximal exercise. *J. Physiol.* **2006**, *574*, 125–138. [CrossRef] [PubMed]

![genes logo] *genes*

MDPI

Article

Inter-Individual Variability in Xenobiotic-Metabolizing Enzymes: Implications for Human Aging and Longevity

Paolina Crocco [†], Alberto Montesanto [†], Serena Dato [†], Silvana Geracitano, Francesca Iannone, Giuseppe Passarino * and Giuseppina Rose *

Department of Biology, Ecology and Earth Sciences, University of Calabria, 87036 Rende, Italy;
crocco.paola@gmail.com (P.C.); alberto.montesanto@unical.it (A.M.); serena.dato@unical.it (S.D.);
silvana.geracitano@unical.it (S.G.); francescaiannonebio@gmail.com (F.I.)
* Correspondence: giuseppe.passarino@unical.it (G.P.); pina.rose@unical.it (G.R.); Tel.: +39-0984-492932 (G.P.);
+39-0984-492931 (G.R.)
† These authors equally contributed.

Received: 29 April 2019; Accepted: 23 May 2019; Published: 27 May 2019

Abstract: Xenobiotic-metabolizing enzymes (XME) mediate the body's response to potentially harmful compounds of exogenous/endogenous origin to which individuals are exposed during their lifetime. Aging adversely affects such responses, making the elderly more susceptible to toxics. Of note, XME genetic variability was found to impact the ability to cope with xenobiotics and, consequently, disease predisposition. We hypothesized that the variability of these genes influencing the interaction with the exposome could affect the individual chance of becoming long-lived. We tested this hypothesis by screening a cohort of 1112 individuals aged 20–108 years for 35 variants in 23 XME genes. Four variants in different genes (*CYP2B6*/rs3745274-G/T, *CYP3A5*/rs776746-G/A, *COMT*/rs4680-G/A and *ABCC2*/rs2273697-G/A) differently impacted the longevity phenotype. In particular, the highest impact was observed in the age group 65–89 years, known to have the highest incidence of age-related diseases. In fact, genetic variability of these genes we found to account for 7.7% of the chance to survive beyond the age of 89 years. Results presented herein confirm that XME genes, by mediating the dynamic and the complex gene–environment interactions, can affect the possibility to reach advanced ages, pointing to them as novel genes for future studies on genetic determinants for age-related traits.

Keywords: aging; longevity; survival; SNP; polymorphism; xenobiotic-metabolizing enzymes; xenobiotics

---

1. Introduction

Aging is a complex phenotype responding to a plethora of drivers in which genetic, behavioral, and environmental factors interact with each other. This can be conceptualized in terms of exposome—that is, the totality of exposures to which an individual is subjected throughout a lifetime and how those exposures affect health [1].

The exposome basically includes a wide variety of toxic or potentially harmful compounds of exogenous (environmental pollutants, dietary compounds, drugs) or endogenous (metabolic by-products such as those resulting from inflammation or lipid peroxidation, oxidative stress, infections, gut flora) origin and related biological responses during the life course [2].

The individual ability to properly cope with xenobiotic stress can influence susceptibility to diseases and, thus, the quality and the rate of aging, phenotypes that certainly result from the cumulative experiences over lifespan. Additionally, in all the different theories proposed to explain the aging process, a common denominator remains the progressive decline of the capacity to deal with environmental stressors to which the human body is constantly exposed.

In this scenario, a crucial role can be played by the coordinated activity of cellular mechanisms evolved for reducing the toxicity of endogenous and xenobiotic compounds to which humans are exposed. These mechanisms comprehend a broad range of reactions of detoxification that make harmful compounds less toxic, more hydrophilic, and easier to be excreted. The main effectors of these mechanisms are a large number of enzymes and transporters, collectively referred to as xenobiotic-metabolizing enzymes (XMEs) or drug metabolizing enzymes (DMEs). This process occurs in three phases. Phase I enzymes, such as cytochrome P450s (CYPs), carboxylesterases, and flavin monooxygenases, add reactive groups to the toxin; in phase II, glutathione S-transferases (GST), UDP-glucuronosyltransferases (UGT), catechol-*O*-methyltransferases (COMT), and *N*-acetyltransferases (NAT) conjugate water-soluble groups onto the molecule; in phase III, ATP-binding cassette (ABC) transporter proteins facilitate the export of the conjugate out of cells as well as the import and the efflux of a broad range of substrates [3].

With aging, there is a decline in the ability to mount a robust response to xenobiotic insults. This is somewhat attributed to the age-related reduction in liver mass, which can result in reduced metabolism rates and in the decreased kidney and liver blood flows, which can result in reduced excretion and elimination of xenobiotic and its metabolites [4]. In addition, a reduction in the activity of phase I and II enzymes and the consequent fall in biotransformation capacity have been reported by several authors in both old animals and humans [5–7]. As aging is characterized by an increased prevalence of chronic conditions that require the use of multiple medications, these changes have particular relevance from a clinical point of view, affecting drug effectiveness and toxicity [8]. Moreover, transcriptional profiling has revealed the up-regulation of xenobiotic-metabolizing genes in long-lived mutants across diverse model organisms [9–11], suggesting that the individual ability to modulate xenobiotic responses may either lead to increased risk of diseases and death or favor longevity.

It is also known that the activity of XME proteins is affected by the variability of the corresponding genes, whose polymorphisms can account for the inter-individual variability in both xenobiotic response/toxicity and disease predisposition. In this regard, significant associations of alleles in these genes (especially in phase II genes) with many forms of cancer [12,13] or coronary heart disease [14] were found. Moreover, in testing a sample of individuals of different ages, Ketelslegers et al. [15] found that the prevalence of risk alleles in XME genes decreases with age, suggesting that individuals carrying a higher number of risk alleles show a higher risk of morbidity and mortality for chronic diseases.

Based on all the above, we reasoned that genetic variants of XME genes might affect the chance to live a long life. In order to test this hypothesis, we screened a set of 35 SNPs in 23 XME genes and their association with aging and survival in a cohort of 1112 individuals aged 20–108 years, performing both case-control and prospective cohort analyses.

2. Materials and Methods

2.1. Study Population

The initial dataset included 1112 unrelated individuals (497 men and 615 women) whose ages ranged from 20 to 108 years. All subjects were born in Calabria (Southern Italy), and their Calabrian ancestry was ascertained up to the third generation. Samples were collected within the framework of several and appropriate recruitment campaigns carried out for monitoring the quality of aging in the whole of Calabria, as previously reported [16]. In brief, younger subjects were recruited from students and staff of the University of Calabria; elderly subjects were from people visiting thermal baths, the Academy of the Elderly, or contacted through general physicians. Very old subjects were selected through the population registers and then contacted and invited to join the study. Old and very old subjects underwent a multidimensional geriatric assessment with the aim of collecting clinical history, anthropometric measures, cognitive functioning, functional activity, and physical performance.

White blood cells (WBC) from blood buffy coats were used as sources of DNA, while plasma/sera were used for routine laboratory analyses.

For the analyses, the sample was divided in three specific age classes based on two age thresholds, 65 and 89 years, after which a significant negative change in the slope of the survival curve of the Italian population occurs [17]. Thus, subjects were classified as younger adults (age class S1, 20–64 years; $n = 330$), elderly (age class S2, 65–89 years; $n = 433$), and very old subjects (age class S3, ≥90 years; $n = 349$).

For subjects of the 65- to 89-year-old group, vital status was traced after a mean follow-up time of approximately 10 years through the population registers of the municipalities where the respondents lived.

2.2. Ethic Statement

The study was approved by the Ethical committee of the University of Calabria (Rende, Italy, on 9 September 2004). All the subjects provided written informed consent in accordance with institutional requirements and the Declaration of Helsinki principles.

2.3. Cognitive and Physical Assessments

Cognitive status was assessed by age- and education-adjusted Mini Mental State Examination (MMSE) [18]. The score ranges from 0 to 30, and a score of 23 points or less is usually considered to indicate cognitive impairment. Hand grip (HG) strength was evaluated by using a handheld dynamometer (SMEDLEY's dynamometer TTM) while the subject was sitting with the arm close to the body. Three consecutive measurements were performed with the stronger hand, and the maximum value was used for data analysis. The performance of activities of daily living (ADL) (bathing, dressing, toileting, transfer from bed to chair, and feeding) was assessed using a modification of the Katz Index [19]. Scores were dichotomized as 1 if the subject was able to perform every activity and as 0 otherwise. Depressive symptoms were assessed using the 15-item Geriatric Depression Scale (GDS) [20]. Subjects with GDS scores greater than or equal to 5 were considered to be affected by depressive symptom.

2.4. SNPs Selection and Primer Design for iPLEX TM Assay

A panel of 35 candidate polymorphisms was selected based on known functional consequence (exonic, regulatory regions) or prior associations with disease risk, pathology, or drug response. SNPs were chosen from 23 genes involved in detoxification related pathways of xenobiotic substances. Supplementary Table S1 reports the complete list of assayed SNPs and their basic features. In summary, we selected 6 SNPs in 5 genes of phase I, 11 SNPs in 7 genes of phase II, and 13 SNPs in 7 genes of phase III. Four non-canonical XME genes (indicated as others in Table S1) and relative polymorphisms (4 SNPs) were selected because they have been deeply studied in relation to drug response, and thus likely affect the risk or the clinical evolution of several diseases. All SNPs have a reported minor allele frequency of >0.05 in Europeans. For each polymorphism, PCR and extension primers were designed using Sequenom MassARRAY Assay Designer 3.0 software (Sequenom, San Diego, CA, USA), resulting in a 22 plex and a 13 plex.

2.5. Sequenom Mass Spectrometry Genotyping

First, 2 µL of genomic DNA (5 ng/uL) were PCR-amplified in a 5 µL reaction containing 0.8 µL HPLC grade water, 0.5 µL of 10 × PCR buffer with 20 mM $MgCl_2$, 0.4 µL of 25 mM $MgCl_2$, 0.1 µL of 25 mM dNTP mix, 1 µL of 0.5 µM primer mix, and 0.2 µL Sequenom PCR enzyme. PCR conditions were: an initial cycle at 94 °C for 2 min, 45 cycles at 95 °C for 30 s, 56 °C for 30 s, 72 °C for 60 s, and a final step at 72 °C for 5 min.

Unincorporated dNTPs in the amplification products were dephosphorylated by adding 2 µL of the shrimp alkaline phosphatase (SAP, Sequenom) mix consisting of 1.53 µL of HPLC grade water,

0.17 µL of SAP buffer, and 0.3 µL (0.5 U) of SAP enzyme (Sequenom). Each reaction was incubated at 37 °C for 40 min, and SAP was then heat inactivated at 85 °C for 5 min.

Following SAP treatment, a single base pair extension reaction was performed using Sequenom's iPLEX Gold chemistry, where 2 µL of the iPLEX reaction mix was added to the samples. The reaction mix consisted of 0.62 µL of HPLC grade water, 0.2 µL of iPlex buffer, 0. 2 µL of iPlex terminator mix, 0.94 µL of primer mix, and 0.04 µL of iPlex enzyme. Thermal cycling conditions included an initial cycle at 94 °C for 30 s; 40 cycles at 94 °C for 5 s, [52 °C for 5 s and 80 °C for 5 s (repeat 5 times per cycle)]; and a final step at 72 °C for 3 min. The samples were then resin treated and spotted on a SpectroCHIP using the MassARRAY nanodispenser (Sequenom) and analyzed using the MassARRAY Compact System matrix-assisted laser desorption/ionization-time-of-flight mass spectrometer (MALDI-TOF) (Sequenom). Genotypes were assigned in real time using the MassARRAY SpectroTYPER RT v3.4 software (Sequenom) based on the mass peaks present. All results were manually inspected using the MassARRAY TyperAnalyzer v3.3 software (Sequenom).

2.6. Quality Control

After genotyping, samples were subjected to a battery of quality control (QC) tests. At sample level, subjects with a proportion of missing genotypes higher than 10% were dropped from the analysis. At SNP level, SNPs were excluded if they had a significant deviation from Hardy–Weinberg equilibrium (HWE, $p < 0.05$) in the younger subgroups, a missing frequency (MiF) higher than 20%, and a minor allele frequency (MAF) lower than 1%. See Table S1 for details.

2.7. Statistical Analysis

For each SNP, allele and genotype frequencies were estimated by gene counting from the observed genotypes. HWE was tested by Fisher's exact test. Logistic regression models were used to evaluate the effect of genotypes (independent variables) on the probability of belonging to different age groups (dependent variable). Differences between age groups were tested by comparing two of them at once. Genetic data were coded with respect to a dominant, a recessive, and an additive model of inheritance. Then, for each SNP, the most likely genetic model was estimated on the basis of minimum level of statistical significance (Wald test p-value). In such models, sex was used as a covariate. To capture sex-dependent effects of the analyzed genetic variants, an additional interaction term was also included. Finally, to test whether combinations of SNPs might better differentiate between the different age groups, a multivariate model including the associated SNPs was also fitted. The Nagelkerke index was then used to compare the obtained models.

Linear and logistic regression models were applied to estimate the impact of genetic variability on parameters of cognitive (MMSE, GDS) and physical (HG, ADL) performance, including age, gender, and height as covariates. Continuous and categorical variables were compared by using the independent samples t-test and the X^2 test as appropriate. For evaluating if the effect of the polymorphisms on longevity phenotype also affected the survival patterns of the different genotypes, we performed a longitudinal study after 10 years from the baseline visit. Univariate survival analysis was carried out by the Kaplan–Meier approach, and survival curves were compared by log-rank test. Subjects were considered as censored if they were alive after the follow-up time, and this time was used as censoring data in the survival analyses. Moreover, hazard ratios (HR) and 95% CI were estimated by Cox proportional hazard models using age and gender as confounder variables.

Because this was a hypothesis driven study, a level of significance p-value = 0.05 was considered for each association test without Bonferroni post hoc correction for multiple comparisons.

Statistical analyses were performed using SNPassoc and surv packages of R (http://www.R-project.org/).

3. Results

After quality control checks, there were genotyping data on 27 SNPs in a cohort of 981 individuals aged 20–108 years. Demographic characteristics for the study cohort according to age groups defined in the Materials and Methods section are presented in Table 1.

Table 1. Demographic characteristics for the analyzed cohort according to age group membership.

	Age Class S1	Age Class S2	Age Class S3
n	287	379	315
Male%	43.6	49.1	35.9
Age Range (years) (mean, SE)	20–64 42.7 (0.89)	65–89 73.5 (0.31)	90–108 97.8 (0.75)
ADL (% Disabled)	-	43.0	69.5
GDS (% Depressed)	-	32.3	26.2
HG (Kg; mean, SE)	-	22.2 (0.62)	13.1 (0.43)
MMSE < 23 (%)	-	9.5	66.5

ADL, activity daily living; GDS, Geriatric Depression Scale; HG, hand grip; MMSE, Mini Mental State Examination; SE, standard error.

Four SNPs demonstrated a nominally significant association (p-value < 0.05) in at least one comparison (see Table 2).

Table 2. Multinomial logistic analysis for univariate genetic associations.

		* Comparison 1 (Age Class 2 vs. Age Class 1)		Comparison 2 (Age Class 3 vs. Age Class 1)		Comparison 3 (Age Class 3 vs. Age Class 2)	
		65–89 Years vs. <65 Years		90+ vs. <65 Years		90+ vs. 65–89 Years	
	Gene	OR (95% CI)	*p*-Value	OR (95% CI)	*p*-Value	OR (95% CI)	*p*-Value
rs3745274-G/T	*CYP2B6*	0.97 (0.67–1.40)	0.88	0.54 (0.37–0.80)	0.002	0.56 (0.39–0.80)	0.005
rs776746-G/A	*CYP3A5*	1.20 (0.65–2.22)	0.57	1.97 (1.08–3.56)	0.022	1.66 (0.99–2.79)	0.054
rs4680-G/A	*COMT*	1.67 (1.10–2.54)	0.016	2.43 (1.58–3.73)	<0.001	1.47 (1.10–2.15)	0.046
rs2273697-G/A	*ABCC2*	0.87 (0.61–1.23)	0.44	1.26 (0.89–1.79)	0.18	1.45 (1.03–2.05)	0.030

* In each comparison, the youngest group was considered as the reference category. For both comparisons 1 and 2 (both using the youngest group as reference category), odd ratios (ORs) were obtained directly from the equations included in the models; for comparison 3 (90+ years vs. <65 years), ORs were obtained by difference of equations included in the models.

Two of them (rs3745274-G/T and rs776746-G/A) belonged to genes for phase I enzymes (*CYP2B6* and *CYP3A5*, respectively), one (rs4680-G/A) was within the phase II *COMT* gene, and one (rs2273697-G/A) was within the phase III *ABCC2* gene. The best-fitting genetic model for three of them was dominant, while rs4680-G/A best fit a recessive genetic model. As Figure 1 shows, these variants had different gene frequency trajectories over the three examined age intervals.

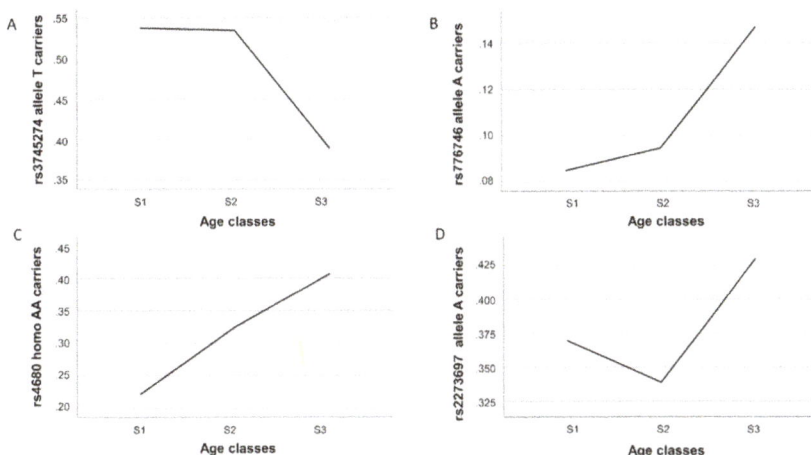

Figure 1. Gene frequencies across the three age classes S1 (20–64 years), S2 (65–89 years), and S3 (90–108 years) of: (**A**) T allele carriers of rs3745274 in *CYP2B6*; (**B**) A allele carriers of rs776746 in *CYP3A5*; (**C**) AA carriers of rs4680 in *COMT*; (**D**) A allele carriers of rs2273697 in *ABCC2*.

We found a significant decrease in the proportion of carriers of the *CYP2B6* rs3745274-T allele in the oldest sample (S3 group) with respect to the youngest S1 and S2 [odds ratio (OR) = 0.547 CI 95% 0.373–0.803, p-value = 0.002 for comparison 2; OR = 0.563 CI 95% 0.395–0.803, p-value = 0.005 for comparison 3), consistent with a detrimental effect of this allele on longevity (Table 2 and Figure 1A). An opposite effect was observed for rs776746, being carriers of the A allele significantly overrepresented in the S3 group as compared to S1 (OR = 1.97 CI 95% 1.08–3.56; p-value = 0.022) and S2 group. In this last case, however, only a trend toward significance was detected (OR = 1.66 CI 95% 0.99–2.79; p-value = 0.054) (Table 2 and Figure 1B). For both SNPs, we did not find differences between age groups in comparison 1, indicating that carrying of the above alleles confers a disadvantageous or an advantageous effect on lifespan only in the last part of life. A different trajectory was observed for the *COMT* rs4680 variant (Figure 1C). Indeed, in all the comparisons, we found that the proportion of homozygous AA individuals was always significantly higher in the oldest than in the younger subjects (OR = 1.67 CI 95% 1.10–2.54, p-value = 0.016 for comparison 1; OR = 2.43 CI 95% 1.58–3.73, p-value <0.001 for comparison 2; OR = 1.47CI 95% 1.10–2.15, p-value = 0.046 for comparison 3), consistent with a linear trend towards a positive effect of the rs4680-AA genotype on longevity. As for the rs2273697 variant in *ABCC2*, a significantly higher prevalence of carriers of the minor allele A in subjects belonging to the S3 group in comparison to the younger age group S2 (comparison 3) (OR = 1.459 CI 95% 1.038–2.051; p-value = 0.030) was observed (Figure 1D), thus indicating a beneficial impact of this allele for reaching longevity. No statistically significant evidence for genotype-by-sex interactions was observed.

Next, to evaluate the overall effect of the multivariate model on the total phenotypic variance, we estimated Nagelkerke indexes for comparisons 1 and 3; we found that the variance explained by the combined genetic data was 0.7% in comparison 1 and 7.7% in comparison 3. Finally, we evaluated the combined effect of the variability of genes listed in Table 2 at different ages. As Table S2 shows, we detected an approximately similar effect size to that seen in univariate analysis, suggesting an independent effect of each SNP. In comparison 3, three out of four SNPs included in the genetic profile remained substantially associated with the phenotype and significantly associated with the age-group membership (rs3745274-G/T, rs776746-G/A, rs4680-G/A), thus significantly discriminating very long lived (90+) from younger elderly (65–89 years old) subjects.

Since the weight of genetic factors increases starting from 65 years of age, we investigated the association of the above variants with biomarkers of age-associated changes in physical (HG and ADL) and cognitive (MMSE and GDS) abilities. A significant association was found between *COMT* rs4680 and ADL performance (*p*-value = 0.03) with subjects homozygous for the allele A showing significantly lower probability to be disabled than those carrying at least one G allele (60.7% of AA among the non-disabled vs. 39.3% of AA among the disabled). In addition, we found that the same variant significantly influenced the GDS performance in females. Subjects with the AA genotype were more represented among non-depressed than depressed individuals (78.9% vs. 51.6%; *p*-value = 0.017). No other significant association between these SNPs and geriatric parameters was observed.

Finally, by using 10-year follow-up survival data, we assessed whether the variants we found associated with the longevity phenotype also influenced the survival of the elderly cohort (age 65–89 years). Kaplan–Meier survival analysis showed a trend for a positive association with survival for carriers of the minor allele (A) of rs2273697 in *ABCC2* (*p*-value = 0.054; see Figure 2), a result consistent with the positive effect on longevity. However, the association did not hold significance when multivariate Cox proportional hazard regression analysis was performed (HR = 0.63, 95% CI: 0.35–1.16; *p*-value = 0.143).

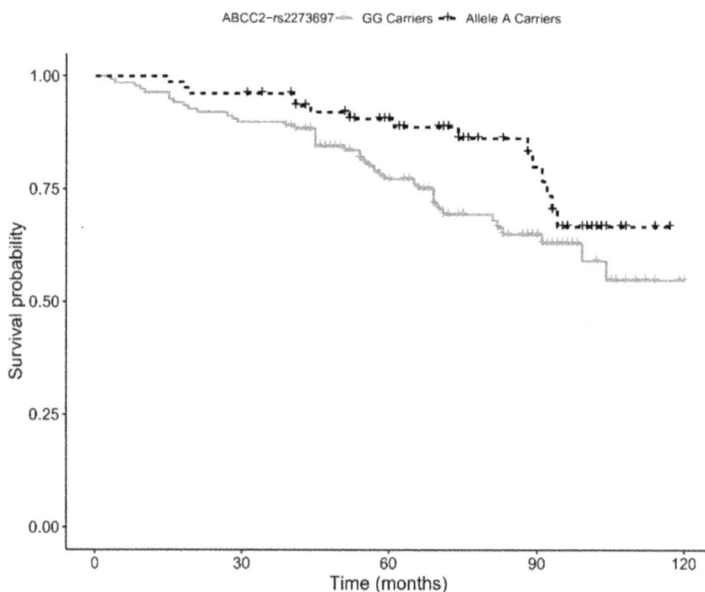

Figure 2. Kaplan–Meier survival functions relative to carriers of the minor allele A (black) vs. non carriers (gray) of the *ABCC2* variant rs2273697. Time is expressed in months, where zero is considered the time of recruitment, and each individual is followed up for survival status until death.

4. Discussion

In this study, we show that genetic variants of genes related to xenobiotic metabolism, such as those of phases I-III, have an influence on the chance of reaching old and very old ages beyond 100 years. Among the 27 genetic variants analyzed, four (rs3745274, rs776746, rs4680, and rs2273697) have shown to exert significantly different and age-specific effects on longevity, with changes of gene frequencies following either linear or non-linear trajectories. In particular, except for rs4680, which showed a significant linear change across the age classes, we observed major frequency changes in the other SNPs in passing from 65–89 to 90–108 age-range. In fact, the genetic variability of these genes showed to account for 7.7% of the chance to survive beyond the age of 89 years. This figure is quite important if we take into account that genetics is believed to account for 25% of the individual chance to be long-lived.

Age-specific gene effects have already been reported in literature for genetic variants in other genes thought to shape the dynamic and the complex gene–environment interactions, which profoundly change during human lifespan [21]. Genes encoding for xenobiotic metabolizing enzymes surely have these characteristics.

Completely opposite effects on longevity were observed for rs3745274-T (negative effect) and rs776746-A carriers (positive effect) located respectively in *CYP2B6* and *CYP3A5* genes that function in biotransformation reactions (phase I). Substrates for both isoenzymes include not only clinically used drugs but also a large number of environmental toxic and carcinogenic chemicals (pollutants, pesticides), as well as endogenous compounds such as steroid hormones and fatty acids [22,23].

The CYP2B6 protein makes up roughly 2–6% of total liver CYP content [24]. There is a remarkable inter-individual variability in its expression partly due to transcriptional regulation and genetic variations [22]. The rs3745274-G/T, herein found to affect the likelihood of becoming long-lived, is a missense variant (Gly516His) in exon 4. A study by Hofmann et al. [25] reported evidence that the T allele is responsible for aberrant splicing, resulting in a shorter variant lacking exon 4, 5, and 6 and decreased expression and enzymatic activity. Notably, relevant studies have established a link between this variant and a higher risk of multiple cancers [26–28]. Therefore, it is likely that individuals carrying the T allele, because of a lower detoxifying capability, have a higher susceptibility to cancer compared to non-carriers and thus a decreased probability of reaching advanced ages.

Emerging evidence also suggests that the rs776746-G/A variant in intron 3 of the *CYP3A5* gene may affect the individual's risk of cancer development. The presence of the G allele creates a cryptic splice site that determines a truncated non-functional protein. Therefore, subjects carrying the GG genotype are considered to be CYP3A5 non-expressors [29]. Very interestingly, the frequency of the rs776746-G allele varies markedly across ethnic groups, ranging from about 18% in Africans to 94% in Europeans (data from 1000 genome) and is significantly correlated with population distance from the equator [30], thus suggesting considerable interaction between genotype and environment. Notably, current literature supports a relationship between the G allele and cancer risk [31,32], suggesting that a protective effect of the expressor A-allele may allow individuals to better cope with dangerous compounds potentially promoting cancer development. Consistently, we found that elderly subjects carrying the rs776746-A have a higher likelihood of becoming long-lived than non-carrying ones. Interestingly, for both rs3745274 and rs776746, changes in gene frequencies start to occur in the middle-aged group (S2), i.e., the one characterized by a higher incidence of cancer and other age-related diseases.

Moreover, the polymorphism rs2273697-G/A appeared to modulate the probability of obtaining longevity by mainly acting in the 65–89 years age group. This is a missense variant (Val417Ile) in exon 10 of the *ABCC2* gene, encoding for the multidrug resistance-associated protein 2 (MRP2), a member of the ABC transporter superfamily. This phase III protein, which is expressed in hepatocytes, renal proximal tubular cells, and enterocytes, is involved in the removal of many toxic chemicals, nutraceuticals, drugs, and their conjugates, as well as endogenous compounds (e.g., bilirubin-glucuronides) [33]. A recent finding by Wei and colleagues provided direct evidence that the rs2273697-A allele increases the ATPase and the efflux activity of the MRP2 protein [34]. Our survival analysis showed that the presence of the A allele decreases the risk of death in subjects aged 65–89 years, and that carriers of this allele are more represented in the 90+ cohort. Based on the above evidence, it seems plausible that rs2273697-A promotes increased survival at old age through an enhanced MRP2 efflux capacity, which likely results in a better protection against cytotoxic effects of toxic compounds.

Finally, we observed a significant linear increase in frequency for the *COMT* rs4680-AA genotype across the different age groups, suggesting a positive effect on survival at both old and very old ages. The rs4680 polymorphism lies in exon 4 and determines either a valine (Val; G allele) or a methionine (Met; A allele) at amino acid 158. This SNP has been reported to significantly affect enzyme activity; for instance, the Val allele has a four-fold higher enzyme activity than the Met allele [35]. COMT is an important phase II enzyme, which, by methylation, inactivates biologically active catechols (i.e., catecholamines, catecholestrogens) and toxic catechol-based molecules [36]. In recent years, COMT has become intensively studied, largely due to its role in regulation of the dopamine level in the brain. In Caucasians, the Met allele is reported to be associated with better cognitive function [37], while Val carriers tend to have a greater likelihood of becoming depressed [38]. It is well known that depression as well as cognition impairment represent major risk factors for disability, often associated with worse health outcomes and increased risk of death, especially in later life [39]. All of the above is in line with our data showing a positive effect on longevity of the A (Met) allele, moreover, confirmed by the association of the same allele with a lower depressive status and a better physical performance in elderly subjects.

On the whole, the different trends in XME gene frequencies we observed in population age once again highlight the complexity in gene–longevity associations. In fact, we found two associated SNPs behaving as pro-longevity variants (rs776746-A and rs4680-AA), one as a killing variant (rs3745274-T), while the rs2273697-A allele showed a U-like frequency curve that was higher at younger ages, decreased in early old, and then increased in exceptionally old. Such a behavior is typical of a buffered variant, in accordance with the buffering mechanism in aging hypothesis suggested by Bergman [40], which states that a deleterious variant can be neutralized by the protective effect of pro-longevity genes.

5. Conclusions

The present study—the first to our knowledge to investigate the association between SNPs in XME genes and human longevity—found that their variability conditions the chance to reach very old age by affecting survival in an age-specific way. This is a novel finding considering that the variability of XME genes has been extensively investigated in relation to drug metabolism and response to treatment. However, drugs are only a limited kind of substrate that XME proteins can metabolize; their detoxification function might rather modulate the dynamic and the complex biological response to the exposome, thus representing a potential determinant of longevity. We are aware that the study is not conclusive and deserves future investigations. Because the genetic variability of all the XME genes herein analyzed show population specificity, it could be very interesting to test the association with longevity in other ethnic groups. Moreover, a longer follow up time and the knowledge of specific causes of death could further support our conclusions, allowing us to better understand the specific pathological phenotype affected by the variants analyzed in this study. Nevertheless, the associations here reported may contribute to our understanding of the genetic determinants of human longevity, supporting future studies in the role of xenobiotic metabolism in quality of aging and extreme survival.

Supplementary Materials: The following are available online at http://www.mdpi.com/2073-4425/10/5/403/s1, Table S1: List of genes and SNPs considered in the present study. Table S2. Multinomial logistic analysis for multivariate genetic associations.

Author Contributions: Conceptualization, G.R., A.M., P.C. and S.D.; methodology, P.C., S.G. and F.I.; formal analysis, A.M.; data curation, A.M. and S.G.; writing—original draft preparation, G.R.; writing—review and editing, G.R., S.D., G.P., P.C. and A.M.; funding acquisition, G.R.

Funding: This research was funded by the Italian Ministry of University and Research (PRIN: Progetti di Ricerca di rilevante Interesse Nazionale—2015, Prot. 20157ATSLF) to G.R.

Acknowledgments: The work has been made possible by the collaboration with Gruppo Baffa (Sadel Spa, Sadel San Teodoro srl, Sadel CS srl, Casa di Cura Madonna dello Scoglio, AGI srl, Casa di Cura Villa del Rosario srl, Savelli Hospital srl, Casa di Cura Villa Ermelinda).

Conflicts of Interest: The authors declare no conflict of interest.

References

1. Wild, C.P. Complementing the genome with an "exposome": The outstanding challenge of environmental exposure measurement in molecular epidemiology. *Cancer Epidemiol. Biomark. Prev.* **2005**, *14*, 1847–1850. [CrossRef] [PubMed]

2. Escher, B.I.; Hackermüller, J.; Polte, T.; Scholz, S.; Aigner, A.; Altenburger, R.; Böhme, A.; Bopp, S.K.; Brack, W.; Busch, W.; et al. From the exposome to mechanistic understanding of chemical-induced adverse effects. *Environ. Int.* **2017**, *99*, 97–106. [PubMed]

3. Xu, C.; Li, C.Y.; Kong, A.N. Induction of phase I, II and III drug metabolism/transport by xenobiotics. *Arch. Pharm. Res.* **2005**, *28*, 249–268. [CrossRef]

4. Klotz, U. Pharmacokinetics and drug metabolism in the elderly. *Drug Metab. Rev.* **2009**, *41*, 67–76. [CrossRef] [PubMed]

5. Wauthier, V.; Verbeeck, R.K.; Calderon, P.B. The effect of ageing on cytochrome P450 enzymes: Consequences for drug biotransformation in the elderly. *Curr. Med. Chem.* **2007**, *14*, 745–757. [CrossRef]

6. Yun, K.U.; Oh, S.J.; Oh, J.M.; Kang, K.W.; Myung, C.S.; Song, G.Y.; Kim, B.H.; Kim, S.K. Age-related changes in hepatic expression and activity of cytochrome P450 in male rats. *Arch. Toxicol.* **2010**, *84*, 939–946. [CrossRef] [PubMed]

7. Vyskočilová, E.; Szotáková, B.; Skálová, L.; Bártíková, H.; Hlaváčová, J.; Boušová, I. Age-related changes in hepatic activity and expression of detoxification enzymes in male rats. *BioMed Res. Int.* **2013**, *2013*, 408573. [CrossRef] [PubMed]

8. Corsonello, A.; Pedone, C.; Incalzi, R.A. Age-related pharmacokinetic and pharmacodynamic changes and related risk of adverse drug reactions. *Curr. Med. Chem.* **2010**, *17*, 571–584. [CrossRef] [PubMed]

9. McElwee, J.J.; Schuster, E.; Blanc, E.; Thomas, J.H.; Gems, D. Shared transcriptional signature in *Caenorhabditis elegans* Dauer larvae and long-lived *daf-2* mutants implicates detoxification system in longevity assurance. *J. Biol. Chem.* **2004**, *279*, 44533–44543. [CrossRef]

10. Gems, D.; McElwee, J.J. Broad spectrum detoxification: The major longevity assurance process regulated by insulin/IGF-1 signaling? *Mech. Ageing Dev.* **2005**, *126*, 381–387. [CrossRef] [PubMed]

11. Amador-Noguez, D.; Dean, A.; Huang, W.; Setchell, K.; Moore, D.; Darlington, G. Alterations in xenobiotic metabolism in the long-lived little mice. *Aging Cell* **2007**, *6*, 453–470. [CrossRef] [PubMed]

12. Hashibe, M.; Brennan, P.; Strange, R.C.; Bhisey, R.; Cascorbi, I.; Lazarus, P.; Ophuis, M.B.; Benhamou, S.; Foulkes, W.D.; Katoh, T.; et al. Meta- and pooled analyses of *GSTM1*, *GSTT1*, *GSTP1*, and *CYP1A1* genotypes and risk of head and neck cancer. *Cancer Epidemiol. Biomark. Prev.* **2003**, *12*, 1509–1517.

13. Lee, K.M.; Kang, D.; Clapper, M.L.; Ingelman-Sundberg, M.; Ono-Kihara, M.; Kiyohara, C.; Min, S.; Lan, Q.; Le Marchand, L.; Lin, P.; et al. *CYP1A1*, *GSTM1*, and *GSTT1* polymorphisms, smoking, and lung cancer risk in a pooled analysis among Asian populations. *Cancer Epidemiol. Biomark. Prev.* **2008**, *17*, 1120–1126. [CrossRef] [PubMed]

14. Izzotti, A.; Cartiglia, C.; Lewtas, J.; De, F.S. Increased DNA alterations in atherosclerotic lesions of individuals lacking the *GSTM1* genotype. *FASEB J.* **2001**, *15*, 752–757. [CrossRef] [PubMed]

15. Ketelslegers, H.B.; Godschalk, R.W.; Gottschalk, R.W.; Knaapen, A.M.; Koppen, G.; Schoeters, G.; Baeyens, W.F.; Nelen, V.; Geraedts, J.P.; van Delft, J.H.; et al. Prevalence of at-risk genotypes for genotoxic effects decreases with age in a randomly selected population in Flanders: A cross sectional study. *Environ. Health* **2011**, *10*, 85. [CrossRef]

16. De Rango, F.; Montesanto, A.; Berardelli, M.; Mazzei, B.; Mari, V.; Lattanzio, F.; Corsonello, A.; Passarino, G. To grow old in southern Italy: A comprehensive description of the old and oldest old in Calabria. *Gerontology* **2011**, *57*, 327–334. [CrossRef]

17. Passarino, G.; Montesanto, A.; Dato, S.; Giordano, S.; Domma, F.; Mari, V.; Feraco, E.; De Benedictis, G. Sex and age specificity of susceptibility genes modulating survival at old age. *Hum. Hered.* **2006**, *62*, 213–220. [CrossRef] [PubMed]

18. Folstein, M.F.; Folstein, S.E.; McHugh, P.R. "Mini-mental state". A practical method for grading the cognitive state of patients for the clinician. *J. Psychiatr. Res.* **1975**, *12*, 189–198. [CrossRef]

19. Katz, S.; Ford, A.B.; Moskowitz, R.W.; Jackson, B.A.; Jaffe, M.W. Studies of illness in the aged. The index of ADL: A standardized measure of biological and psychosocial function. *JAMA* **1963**, *185*, 914–919. [CrossRef] [PubMed]

20. Lesher, E.L.; Berryhill, J.S. Validation of the Geriatric Depression Scale—Short Form among inpatients. *J. Clin. Psychol.* **1994**, *50*, 256–260. [CrossRef]

21. Giuliani, C.; Sazzini, M.; Pirazzini, C.; Bacalini, M.G.; Marasco, E.; Ruscone, G.A.G.; Fang, F.; Sarno, S.; Gentilini, D.; Di Blasio, A.M.; et al. Impact of demography and population dynamics on the genetic architecture of human longevity. *Aging (Albany NY)* **2018**, *10*, 1947–1963. [CrossRef] [PubMed]

22. Turpeinen, M.; Zanger, U.M. Cytochrome P450 2B6: Function, genetics, and clinical relevance. *Drug Metabol. Drug Interact.* **2012**, *27*, 185–197. [CrossRef] [PubMed]

23. Lamba, J.; Hebert, J.M.; Schuetz, E.G.; Klein, T.E.; Altman, R.B. PharmGKB summary: Very important pharmacogene information for *CYP3A5*. *Pharmacogenet. Genom.* **2012**, *22*, 555–558. [CrossRef] [PubMed]

24. Zanger, U.M.; Klein, K. Pharmacogenetics of cytochrome P450 2B6 (CYP2B6): Advances on polymorphisms, mechanisms, and clinical relevance. *Front. Genet.* **2013**, *4*, 24. [CrossRef]

25. Hofmann, M.H.; Blievernicht, J.K.; Klein, K.; Saussele, T.; Schaeffeler, E.; Schwab, M.; Zanger, U.M. Aberrant splicing caused by single nucleotide polymorphism c.516G>T [Q172H], a marker of *CYP2B6*6*, is responsible for decreased expression and activity of CYP2B6 in liver. *J. Pharmacol. Exp. Ther.* **2008**, *325*, 284–292. [CrossRef] [PubMed]

26. Yuan, Z.H.; Liu, Q.; Zhang, Y.; Liu, H.X.; Zhao, J.; Zhu, P. *CYP2B6* gene single nucleotide polymorphisms and leukemia susceptibility. *Ann. Hematol.* **2011**, *90*, 293–299. [CrossRef]

27. Justenhoven, C.; Pentimalli, D.; Rabstein, S.; Harth, V.; Lotz, A.; Pesch, B.; Brüning, T.; Dörk, T.; Schürmann, P.; Bogdanova, N.; et al. *CYP2B6*6* is associated with increased breast cancer risk. *Int. J. Cancer* **2014**, *134*, 426–430. [CrossRef]

28. Daraki, A.; Zachaki, S.; Koromila, T.; Diamantopoulou, P.; Pantelias, G.E.; Sambani, C.; Aleporou, V.; Kollia, P.; Manola, K.N. The $G^{516}T$ *CYP2B6* germline polymorphism affects the risk of acute myeloid leukemia and is associated with specific chromosomal abnormalities. *PLoS ONE* **2014**, *9*, e88879. [CrossRef]

29. Perera, M.A. The missing linkage: What pharmacogenetic associations are left to find in CYP3A? *Expert Opin. Drug Metab. Toxicol.* **2010**, *6*, 17–28. [CrossRef]

30. Thompson, E.E.; Kuttab-Boulos, H.; Witonsky, D.; Yang, L.; Roe, B.A.; Di Rienzo, A. CYP3A variation and the evolution of salt-sensitivity variants. *Am. J. Hum. Genet.* **2004**, *75*, 1059–1069. [CrossRef]

31. Wang, B.S.; Liu, Z.; Xu, W.X.; Sun, S.L. *CYP3A5*3* polymorphism and cancer risk: A meta-analysis and meta-regression. *Tumor Biol.* **2013**, *34*, 2357–2366. [CrossRef]

32. Liang, Y.; Han, W.; Yan, H.; Mao, Q. Association of *CYP3A5*3* polymorphisms and prostate cancer risk: A meta-analysis. *J. Cancer Res. Ther.* **2018**, *14*, S463–S467.

33. Jemnitz, K.; Heredi-Szabo, K.; Janossy, J.; Ioja, E.; Vereczkey, L.; Krajcsi, P. ABCC2/Abcc2: A multispecific transporter with dominant excretory functions. *Drug Metab. Rev.* **2010**, *42*, 402–436. [CrossRef] [PubMed]

34. Wei, D.; Zhang, H.; Peng, R.; Huang, C.; Bai, R. ABCC2 (1249G > A) polymorphism implicates altered transport activity for sorafenib. *Xenobiotica* **2017**, *47*, 1008–1014. [CrossRef]

35. Chen, J.; Lipska, B.K.; Halim, N.; Ma, Q.D.; Matsumoto, M.; Melhem, S.; Kolachana, B.S.; Hyde, T.M.; Herman, M.M.; Apud, J.; et al. Functional analysis of genetic variation in catechol-O-methyltransferase (COMT): Effects on mRNA, protein, and enzyme activity in postmortem human brain. *Am. J. Hum. Genet.* **2004**, *75*, 807–821. [CrossRef]

36. Männistö, P.T.; Kaakkola, S. Catechol-O-methyltransferase (COMT): Biochemistry, molecular biology, pharmacology, and clinical efficacy of the new selective COMT inhibitors. *Pharmacol. Rev.* **1999**, *51*, 593–628.

37. Aguilera, M.; Barrantes-Vidal, N.; Arias, B.; Moya, J.; Villa, H.; Ibanez, M.I.; Ruiperez, M.A.; Ortet, G.; Fananas, L. Putative role of the *COMT* gene polymorphism (Val158Met) on verbal working memory functioning in a healthy population. *Am. J. Med. Genet. B Neuropsychiatr. Genet.* **2008**, *147B*, 898–902. [CrossRef] [PubMed]

38. Wang, M.; Ma, Y.; Yuan, W.; Su, K.; Li, M.D. Meta-analysis of the *COMT* Val158Met polymorphism in major depressive disorder: effect of ethnicity. *J. Neuroimmune Pharmacol.* **2016**, *11*, 434–445. [CrossRef] [PubMed]

39. Mehta, K.M.; Yaffe, K.; Langa, K.M.; Sands, L.; Whooley, M.A.; Covinsky, K.E. Additive effects of cognitive function and depressive symptoms on mortality in elderly community-living adults. *J. Gerontol. A Biol. Sci. Med. Sci.* **2003**, *58*, 461–467. [CrossRef]

40. Bergman, A.; Atzmon, G.; Ye, K.; MacCarthy, T.; Barzilai, N. Buffering mechanisms in aging: A systems approach toward uncovering the genetic component of aging. *PLoS Comput. Biol.* **2007**, *3*, e170. [CrossRef]

![genes](GCAT TACG GCAT) *genes*

MDPI

Article

Exceptional Longevity and Polygenic Risk for Cardiovascular Health

Mary Revelas [1,2], Anbupalam Thalamuthu [1,2], Christopher Oldmeadow [3], Tiffany-Jane Evans [3], Nicola J. Armstrong [1,4], Carlos Riveros [3,5], John B. Kwok [2,6], Peter R. Schofield [2,6], Henry Brodaty [1,7], Rodney J. Scott [5,8], John R. Attia [3,8], Perminder S. Sachdev [1,9] and Karen A. Mather [1,2,*]

[1] Centre for Healthy Brain Ageing, School of Psychiatry, UNSW Medicine, University of New South Wales, Sydney, NSW 2031, Australia; maryg922@gmail.com (M.R.); a.thalamuthu@unsw.edu.au (A.T.); N.Armstrong@murdoch.edu.au (N.J.A.); h.brodaty@unsw.edu.au (H.B.); p.sachdev@unsw.edu.au (P.S.S.)
[2] Neuroscience Research Australia, Randwick, NSW 2031, Australia; john.kwok@sydney.edu.au (J.B.K.); p.schofield@neura.edu.au (P.R.S.)
[3] Hunter Medical Research Institute, Newcastle, NSW 2305, Australia; christopher.oldmeadow@newcastle.edu.au (C.O.); tiffany.evans@hmri.org.au (T.-J.E.); carlos.riveros@hmri.org.au (C.R.); John.Attia@newcastle.edu.au (J.R.A.)
[4] Discipline of Mathematics and Statistics, Murdoch University, Perth, WA 6150, Australia
[5] Faculty of Health, University of Newcastle, Newcastle, NSW 2308, Australia; rodney.scott@newcastle.edu.au
[6] School of Medical Sciences, University of New South Wales, Sydney, NSW 2052, Australia
[7] Dementia Centre for Research Collaboration, University of New South Wales, Sydney, NSW 2052, Australia
[8] Pathology North, John Hunter Hospital, Newcastle, NSW 2305, Australia
[9] Neuropsychiatric Institute, Prince of Wales Hospital, Barker Street, Randwick, NSW 2031, Australia
* Correspondence: Karen.mather@unsw.edu.au; Tel.: +61-(2)-9399-1064

Received: 31 January 2019; Accepted: 14 March 2019; Published: 18 March 2019

Abstract: Studies investigating exceptionally long-lived (ELL) individuals, including genetic studies, have linked cardiovascular-related pathways, particularly lipid and cholesterol homeostasis, with longevity. This study explored the genetic profiles of ELL individuals (cases: n = 294, 95–106 years; controls: n = 1105, 55–65 years) by assessing their polygenic risk scores (PRS) based on a genome wide association study (GWAS) threshold of $p < 5 \times 10^{-5}$. PRS were constructed using GWAS summary data from two exceptional longevity (EL) analyses and eight cardiovascular-related risk factors (lipids) and disease (myocardial infarction, coronary artery disease, stroke) analyses. A higher genetic risk for exceptional longevity (EL) was significantly associated with longevity in our sample (odds ratio (OR) = 1.19–1.20, p = 0.00804 and 0.00758, respectively). Two cardiovascular health PRS were nominally significant with longevity (HDL cholesterol, triglycerides), with higher PRS associated with EL, but these relationships did not survive correction for multiple testing. In conclusion, ELL individuals did not have significantly lower polygenic risk for the majority of the investigated cardiovascular health traits. Future work in larger cohorts is required to further explore the role of cardiovascular-related genetic variants in EL.

Keywords: polygenic risk score; cardiovascular health; exceptional longevity; lipid profile

---

1. Introduction

The human life span has significantly increased over the last century with many individuals surpassing 80 years of age in developed countries due to factors such as improved healthcare and favourable lifestyle choices [1]. Exceptional longevity, defined as exceeding the average life expectancy, is multifaceted with genetic, environmental and epigenetic factors all playing a role. Exceptionally long-lived (ELL) individuals are examples of successful ageing with a proportion demonstrating compression of morbidity [2]. Thus, ELL individuals have been described as "super controls" for studies on age-related decline and disease [3]. It is important to study these models of successful ageing, as these rare individuals may reveal novel longevity-associated pathways, which may ultimately translate into strategies to promote health in our ageing population.

There is evidence linking healthier cardiovascular risk profiles and lower incidence of cardiovascular disease with longevity [4]. Analysis of lipid metabolism in longevous families identified changes in lipid concentration, specifically a smaller total cholesterol to high-density lipoprotein-cholesterol (TC/HDL-C) ratio and lower triglycerides levels, in the offspring of nonagenarians [5]. Lipid profiling in the Leiden Longevity Study established larger low-density lipoprotein particles as major predictors of longevity [5]. Similarly, Barzilai et al. (2015) suggested that healthy ageing is promoted by a unique lipoprotein profile [6]. Levels of apolipoproteins, important lipid transporters in the circulatory system, have been observed to decline with age. However, higher apolipoprotein levels in the exceptionally long lived have been reported, suggesting a younger apolipoprotein profile that may promote longevity [7].

Further evidence from candidate and genome-wide association longevity studies indicates that cardiovascular pathways are involved in successful ageing. In a parental longevity genome wide association study (GWAS) examining 75,000 participants, the authors concluded that cardiovascular-related pathways are important contributors in attaining exceptional longevity [8]. In a recent meta-analysis examining longevity genetic polymorphisms, all significant genes (*APOE*, *FOXO3A*, *ACE*, *Klotho* and *IL6*) play roles in cardiovascular pathways, such as lipid metabolism, or have been previously linked to cardiovascular disease [9]. The genetic variants with the largest effect sizes in this meta-analysis were located in the *APOE* and *FOXO3A* genes. The product of the *APOE* gene transports lipids in the blood and is consequently critical in cholesterol metabolism. Any loss of cholesterol homeostasis may increase the risk of cardiovascular disease, obesity and diabetes [10]. In addition, cardiovascular diseases (heart attack and stroke) have been strongly linked to *APOE* [11]. *FOXO3A* is an evolutionary conserved transcription factor and has been consistently and independently replicated with longevity in many ethnically diverse cohorts [12]. Recently, the longevity-associated G allele of *FOXO3* rs2802292 in older Japanese and Caucasians was associated with decreased risk of coronary artery disease mortality [13]. Interestingly, both Ashkenazi Jewish and Italian centenarians genotyped for an isoleucine to valine variation at codon 405 in the cholesteryl ester transfer protein (*CETP*) gene had a higher frequency of the VV genotype, which has been associated with larger low-density lipoprotein particle sizes and lower *CETP* serum levels. The authors deduced that lipoprotein particle size is heritable and encourages healthy aging [6,14]. Similarly, homozygosity for the -641 C allele in the *APOC3* promoter (rs2542052) was 25% higher in centenarians and linked with pro-longevity lipoprotein levels and sizes [15]. Additional prior studies investigating genetic vascular factors involved in human longevity have described variants in genes involved in blood pressure regulation such as methyltetrahydrofolatereductase (*MTHFR*), paraoxonase 1 (*PON1*) and plasminogen activator inhibitor type I (*PAI-1*) [16]; however, further studies are still needed to confirm these suggested associations.

Polygenic risk scores (PRS) for cardiovascular-related phenotypes can now be calculated due to the availability of summary data from GWAS examining a broad range of traits from lipids to coronary artery disease. This facilitates the evaluation of the contribution of polygenic risk for cardiovascular risk factors and disease to exceptional longevity and successful ageing. Thus, the purpose of this study was to explore the genetic profiles of ELL individuals aged (≥95 years) by assessing their polygenic risk for cardiovascular-related risk and disease phenotypes relative to middle-aged controls. This study tests the hypothesis that ELL individuals have lower polygenic risk for cardiovascular health-related traits and disease compared to controls, which may give them a survival advantage.

2. Materials and Methods

2.1. Participants

ELL individuals were recruited from two Sydney-based Australian studies: The Sydney Centenarian Study (SCS) [17] and the Sydney Memory and Ageing Study (Sydney MAS) [18]. Both studies recruited participants using the compulsory electoral roll and Medicare lists from New South Wales, Australia. A subsample from SCS with available genetic data provided 256 long-lived cases with a European background (age range 95–106, mean age 97.5 years, 31% male). This sample was enriched by the addition of 38 individuals from Sydney MAS (≥95 years, mean age 86.7 years, 26% male). Ethics approval was granted from the relevant Human Research Ethics Committees for each study. SCS and Sydney MAS were approved by the Human Research Ethics Committees of UNSW Sydney and the South Eastern Sydney and Illawarra Area Health Service (ethics approvals HC17251 and HC14327, respectively). The Hunter Community Study (HCS) was approved by the University of Newcastle and Hunter New England Human Research Ethics Committees (HREC 03/12/10/3.26). Written informed consent was obtained from all participants or if unable to consent, proxy consent was obtained from the nearest of kin.

Controls were obtained from the Hunter Community Study (HCS) in Newcastle, New South Wales [19]. These participants have a similar ethnic background to the long-lived cases and Newcastle is geographically close to Sydney (160 km). The HCS is a cohort of 3253 individuals (age range 55–85, mean age 66.3 years, 46% male). For the purpose of this investigation a subsample of 1105 individuals aged 55–65 (mean age 60.3 years, 47% male) were used as controls.

Fasting blood samples from each cohort were collected for DNA and biochemistry analyses. Biological measures assessed included total cholesterol, as well as low-density lipids (LDL) and high-density lipids (HDL), and triglycerides as described in [20,21].

2.2. Genotyping

DNA was extracted using standard methods. SCS cases were genotyped using the Illumina OmniExpress array (California, USA), whereas the HCS cases were genotyped using the Affymetrix Axiom Kaiser array (California, USA) and the Sydney MAS samples were genotyped using the Affymetrix Genome-wide Human SNP Array 6.0 according to the manufacturer's instructions. In all three cohorts, genotyped single nucleotide polymorphisms (SNPs) were excluded if the following criteria were observed: (i) The call rate was <95%, (ii) p-value for Hardy-Weinberg equilibrium was <10^{-6}, (iii) minor allele frequency was <0.01% and (iv) the strand ambiguous (A/T and C/G). If first- or second-degree relatives were identified, only one family member was retained for analysis. EIGENSTRAT analysis [22] allowed for the detection and removal of any ethnic outliers. After quality control (QC) checks, for SCS and Sydney MAS there were genotyping data on 640,355 and 734,550 SNPs, respectively, whilst for HCS there were data on 739,276 SNPs. As part of the QC checks, the reported sex of the participants was verified using genotyped data and samples with sex discrepancies were discarded. The quality controlled genotype data were imputed in the Michigan imputation server (https://imputationserver.sph.umich.edu) [23] using the Haplotype Reference Consortium reference panel (v3.20101123). Similar to the genotyped data, QC steps in the imputed data were implemented (minor allele frequency (MAF) > 0.05, imputation quality score >0.6, call rate >0.95, HWE p-value >10^{-6}). The SNPs with high quality dosage scores were converted to best-guess genotypes using PLINK and were used in the calculation of the PRS. The genotypes for the *APOE* single nucleotide polymorphisms (SNPs) rs7412 and rs429358 were extracted from the imputed dosage using PLINK. Both the SNPs were imputed with high accuracy in all the three cohorts (r2 > 0.80) and the *APOE* ε2/3/4 haplotypes were inferred using these two SNPs [24].

Genotyping availability: Due to ethical concerns, genotyping data have not been deposited in a public repository. However, genotyping data can be requested via a formal review process to the relevant studies.

2.3. Polygenic Risk Scores (PRS)

PRS were generated by using the PRSice program [25] from summary statistics obtained from previous GWAS studies. The following phenotypes were examined: Longevity, cardiovascular disease (myocardial infarction, all stroke and coronary artery disease) and cardiovascular disease-related risk factors (cholesterol, triglycerides, high- and low-density lipoproteins and essential hypertension). Details regarding each GWAS utilised, including links to the summary data are provided in the Supplementary (Tables S1 and S2). Linkage disequilibrium pruning was performed using the clumping option (r2 > 0.25 and physical distance threshold of 250 kb KB). PRS were calculated for each of the phenotypes using different *p*-value significance cut-offs for the SNPs included from each of the GWAS, ranging from $p = 5 \times 10^{-8}$ to 1. We only present the results for the PRS *p*-value threshold of 5×10^{-5} in the main text. Due to the strong associations between longevity and (i) the *APOE* locus (chromosome 19, base pair 45,393,826–45,422,606) and (ii) the *FOXO3A* locus (chromosome 6, base pair 108,881,038–109,005,977), longevity PRS were also calculated after removing these loci.

2.4. Statistical Analyses

All analyses were performed using R version 3.4.3 [26]. Chi-squared tests were utilised to determine differences in proportions between groups. To achieve normality of the lipid variables, inverse normal transformations were performed. Independent sample t tests were used to compare the means between the cases and controls.

To demonstrate that the PRS were associated with their respective phenotypes in our sample, analyses were undertaken investigating the relationships between (a) PRS for EL phenotypes (+/−*APOE*, *FOXO3A* and *APOC3* loci) and EL in our sample using logistic regression, controlling for sex and where appropriate *APOE* ε4 or *APOC3* (rs2542052) C homozygotes; (b) cardiovascular risk factor (lipids) PRS and measured lipid levels using linear regression, adjusting for age and sex and where appropriate *APOE* ε4 or *APOC3* (rs2542052) C homozygotes. Lastly, logistic regressions were used to investigate associations between cardiovascular health PRS and EL in our sample, controlling for sex.

The variance explained by the PRS using logistic regression was calculated using the Nagelkerke method [27] as implemented in the R package rsq [28] and power calculation was performed using WebPower [29]. Correlation analyses between the different PRS were undertaken to examine genetic overlap, using the PRS calculated at the nominated GWAS threshold *p*-value cut-off of 5×10^{-5}.

3. Results

3.1. Sample Characteristics

The sample characteristics are described in Table 1. As expected, there were fewer *APOE* ε4 carriers in the ELL cases compared to the younger controls, 14.6% versus 30.8%, respectively ($p = 1.49 \times 10^{-8}$). LDL and total cholesterol levels were statistically different, with controls displaying higher lipid levels than EL cases (Table 1). The frequency of *APOC3* (rs2542052) C homozygotes did not differ significantly in the cases and the controls ($p = 0.503564$).

Table 1. Sample characteristics of long-lived cases (≥95 years) and younger controls (55–65 years).

	Cases	Controls	*p*-Value
Cohort	SCS/Sydney MAS	HCS	
Sample size (n)	294	1105	N/A
Age range (mean ± SD)	95–106 (96.1 ± 4.1)	55–65 (60.3 ± 2.8)	N/A
N (%) males	90 (31)	518 (47)	N/A
APOC3 [a] C homozygotes, n (%)	120 (40.1)	475 (42.9)	0.503564
APOE ε4 carrier, n (%)	43 (14.6)	340 (30.8)	1.49×10^{-8}
HDL (mean ± SD) (% missing)	1.47 ± 0.45 (17.1)	1.36 ± 0.37 (0.6)	0.148557
LDL (mean ± SD) (% missing)	2.69 ± 1.02 (17.3)	3.25 ± 0.91 (12.3)	1.23×10^{-19}
TC (mean ± SD) (% missing)	4.75 ± 1.16 (17.0)	5.24 ± 1.03 (0)	6.37×10^{-15}
TG (mean ± SD) (% missing)	1.31 ± 0.63 (17.0)	1.41 ± 1.12 (0.6)	0.492286

Notes: HDL = high-density lipoprotein, LDL = low-density lipoprotein, TC = total cholesterol, TG = total triglyceride. Raw mean values are reported. [a], *APOC3* SNP rs2542052.

3.2. Polygenic Risk Scores (PRS)

The exceptional longevity and cardiovascular risk and disease phenotypes used for calculation of the PRS are shown in Table 2. The total number of SNPs included from each GWAS for the PRS calculations are listed for two GWAS p-thresholds (genome-wide significance $p < 5 \times 10^{-8}$, suggestive $p < 5 \times 10^{-5}$). Details for all other calculated PRS calculated (at different *p*-value cut-offs) are available in the Supplementary (Table S3).

Table 2. Longevity and cardiovascular-related risk and disease phenotypes examined, with the number of single nucleotide polymorphisms (SNPs) included in each polygenic risk score (PRS) at two GWAS *p*-value cut-offs.

Phenotype [GWAS ref]	N SNPs $p < 5 \times 10^{-8}$	N SNPs $p < 5 \times 10^{-5}$	Total Number of GWAS SNPs Available
Exceptional longevity			
Exceptional longevity [30]	0	32	184,562
Exceptional parental longevity [8]	0	123	476,093
Cardiovascular health			
Myocardial infarction [31]	35	107	19,607
Stroke [32]	11	226	471,632
Coronary artery disease [33]	96	447	460,589
Essential hypertension (http://www.nealelab.is)	2	107	476,069
HDL [34]	318	685	204,118
LDL [34]	301	652	202,316
Cholesterol [34]	367	797	204,123
Triglycerides [34]	238	592	201,879

Notes: SNP numbers are after QC steps for the PRS calculation, including removal of any SNPs with a poor imputation quality score (≤ 0.6) and linkage disequilibrium pruning. HDL = high-density lipoprotein, LDL = low-density lipoprotein.

3.3. PRS Associations with Measured Phenotypes

3.3.1. Longevity PRS with Exceptional Longevity

As expected, higher longevity PRS from either Broer et al. [30] (EL, ≥ 90 cases vs. controls) or Pilling et al. [8] (exceptional parental longevity—EPL cases had a mother who lived ≥ 98 years and a father ≥ 95 years vs. controls) were significantly associated with longevity in our sample (Table 3, Figure 1). Results from both analyses suggest individuals carrying pro-longevity variants were 1.19–1.20 times more likely (per standard deviation increase in PRS) to survive to an exceptional age (≥ 95 years) or to have both parents that were ELL. PRS calculated at other *p*-value thresholds were all statistically significant (Table S4). For example, at a *p*-value threshold of 0.05, the EL PRS had an OR of 1.79 ($p = 2.29 \times 10^{-17}$), and the PRS for exceptional parental longevity had an OR of 2.01 ($p = 6.34 \times 10^{-23}$). Sex was also found to be significant. Figure 1 shows that within the ELL cases, there is a subsample that have low-longevity PRS (i.e., for both EL and EPL) and yet have survived to 95 years old and over. The PRS explained 0.7%–10% of the phenotype variance depending on the GWAS *p*-value threshold (Table S4).

Table 3. Associations between longevity EL and EPL PRS (*p*-value threshold $< 5 \times 10^{-5}$) and exceptionally long-lived cases versus controls.

	Odds Ratio (OR)	Standard Error [8]	*p*-Value
PRSEL [30]	1.20	0.068	0.00758
PRSEPL [8]	1.19	0.067	0.00804

Notes: EPL = exceptional parental longevity, defined as participants with a mother who lived ≥ 98 years and a father ≥ 95 years. EL = exceptional longevity, defined as participants who lived ≥ 90 years. Logistic regressions were adjusted for sex, comparing differences in PRS for ELL vs. controls. ORs expressed in PRS SD units.

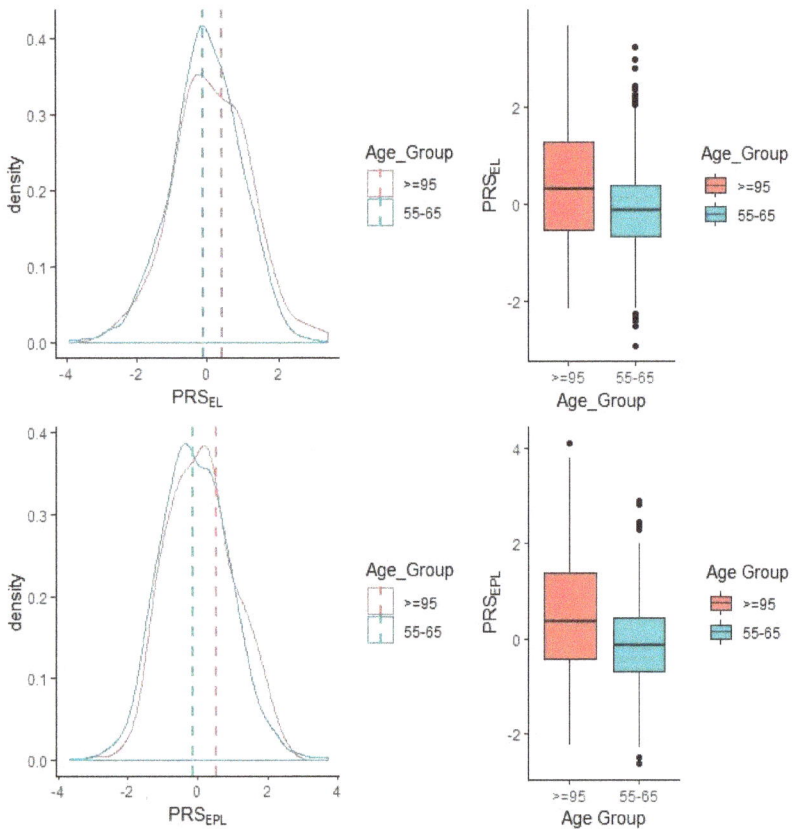

Figure 1. Density plots visualizing the standardised PRS distributions for EL (exceptional longevity, PRSEL) and EPL (exceptional parental longevity, PRSEPL) in the entire sample and the corresponding box plots in the long-lived cases (95+ years) versus controls (55–65 years).

Additional analyses excluding the *TOMM40/APOE/APOC1* locus (Table S5), the *FOXO3A* locus (Table S6) and the *APOC3* locus (Table S7) did not markedly change the results. When adjusting for ε4 carrier status or *APOC3* C homozygotes, the results did not change noticeably (Tables S8 and S9).

3.3.2. Association of Cardiovascular Risk Factor (Lipids) PRS with Measured Lipid Levels

Linear regressions were performed to ascertain whether the PRS for the lipid cardiovascular risk factors were associated with actual lipid levels in the current study. As expected, (Table 4), the PRS were positively associated with the measured lipid levels in our cohorts. The results for all lipid PRS calculated at additional *p*-value thresholds and additionally adjusted for *APOE* ε4 or *APOC3* C homozygotes are available in the Supplementary (Tables S10, S11 and S12, respectively) The additional adjustments did not change the results significantly.

3.3.3. Cardiovascular Health PRS and Exceptional Longevity

As shown in Figure 2, only two cardiovascular health PRS (HDL and TG) at the chosen *p*-value threshold of $< 5 \times 10^{-5}$ were nominally significant with exceptional longevity, with higher PRS associated with exceptional longevity. However, after Bonferroni correction for multiple testing, assuming eight independent tests, they no longer remained significant. The results for all cardiovascular health PRS calculated at additional *p*-value thresholds are available in the Supplementary (Table S13). Analyses were also adjusted for *APOE* ε4 and *APOC3* C homozygotes (Tables S14 and S15; when adjusting for *APOE* ε4, three cardiovascular health PRS (HDL, TG and myocardial infarction (MI)) were marginally significant but would not survive multiple testing correction.

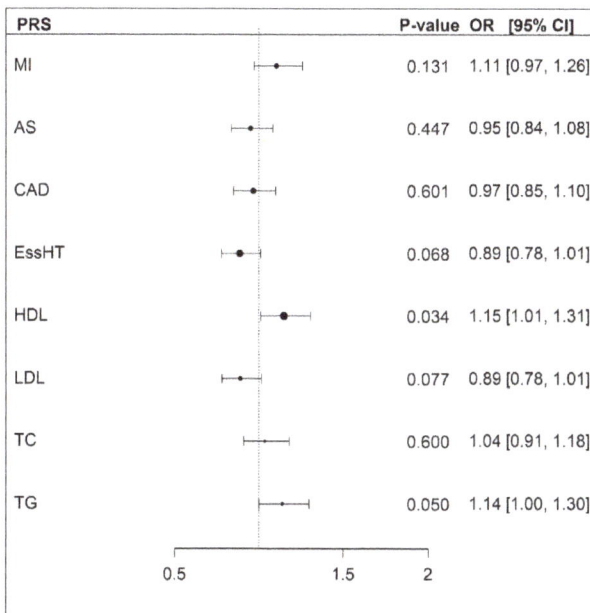

PRS		P-value	OR [95% CI]
MI		0.131	1.11 [0.97, 1.26]
AS		0.447	0.95 [0.84, 1.08]
CAD		0.601	0.97 [0.85, 1.10]
EssHT		0.068	0.89 [0.78, 1.01]
HDL		0.034	1.15 [1.01, 1.31]
LDL		0.077	0.89 [0.78, 1.01]
TC		0.600	1.04 [0.91, 1.18]
TG		0.050	1.14 [1.00, 1.30]

Figure 2. Forest plot showing the associations of different cardiovascular PRS with exceptional longevity. Notes. MI = myocardial infarction [31], AS = all stroke [32], CAD = coronary artery disease [33], EssHT = essential hypertension [www.nealelab.is], HDL = high-density lipoprotein, LDL = low-density lipoprotein, TC = total cholesterol, TG = total triglyceride [34]. Logistic regression analyses were adjusted for sex, comparing differences in PRS for ELL vs. controls. PRS were calculated using the relevant GWAS summary results with a *p*-value threshold $< 5 \times 10^{-5}$.

Table 4. Associations of different PRS (p-value $< 5 \times 10^{-5}$) for lipid cardiovascular risk factors with measured lipid levels in the combined sample (55–106 years).

	Sample Size (*n*)	Beta (β)	Standard Error [8]	*p*-Value
PRSHDL	1331	0.261	0.024	3.94×10^{-26}
PRSLDL	1196	0.200	0.028	3.81×10^{-13}
PRSTC	1336	0.176	0.026	1.66×10^{-11}
PRSTG	1335	0.254	0.026	2.63×10^{-21}

Notes. HDL = high-density lipoprotein, LDL = low-density lipoprotein, TC = total cholesterol, TG = total triglyceride. Linear regression adjusted for age and sex, independent variable = PRS, dependent variable = lipid level. All lipid PRS were generated using summary results from Willer et al. [34].

PRS Genetic Overlap

Figure 3 explores the genetic overlap between the investigated PRS. Exceptional parental longevity overlaps with the exceptional longevity PRS, although the correlation coefficient is small (r = 0.07, *p* = 0.009). *P*-values for the entire correlation matrix are provided in Table S16. Interestingly, LDL overlaps with both exceptional longevity (r = −0.06, *p* = 0.03) and exceptional parental longevity (r = −0.05, p = 0.04) in the expected direction. Myocardial infarction (MI) overlaps with exceptional longevity (r = −0.06, p = 0.03) but not exceptional parental longevity. Other significant correlations were observed between the cardiovascular risk factors and/or cardiovascular disease. For example, the MI PRS was correlated with the PRS for stroke, coronary artery disease, HDL, LDL, total cholesterol and triglycerides.

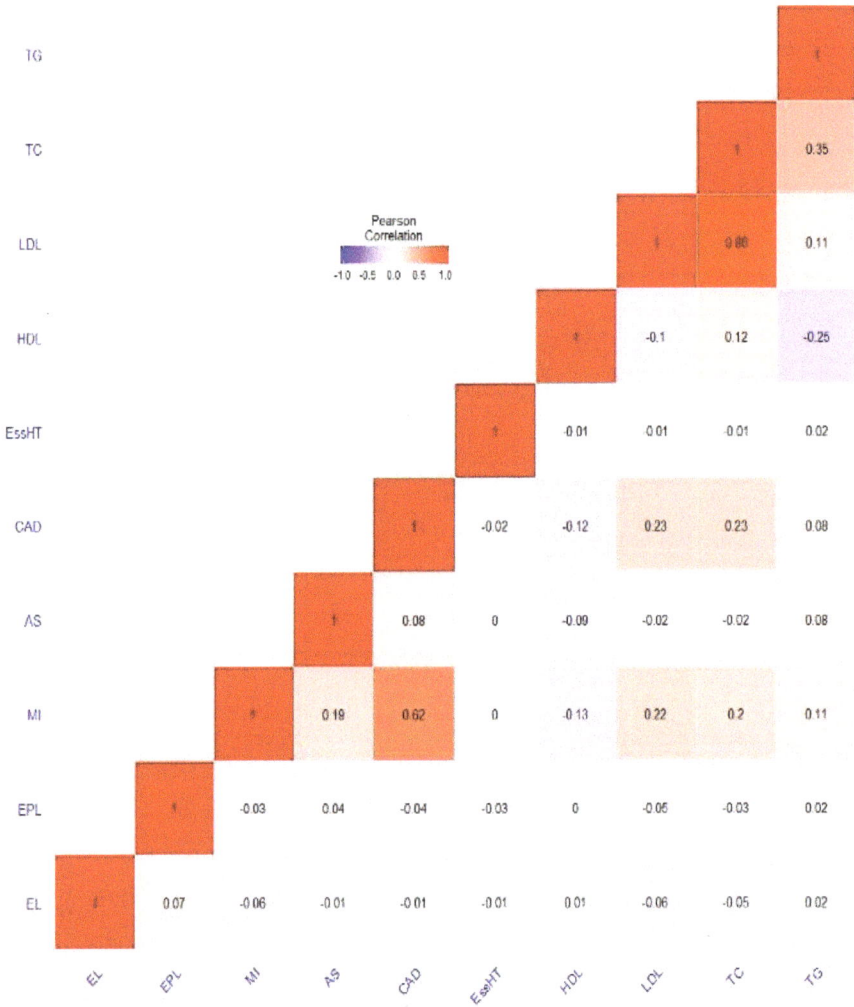

	EL	EPL	MI	AS	CAD	EssHT	HDL	LDL	TC	TG
TG										1
TC									1	0.35
LDL								1	0.88	0.11
HDL							1	-0.1	0.12	-0.25
EssHT						1	-0.01	0.01	-0.01	0.02
CAD					1	-0.02	-0.12	0.23	0.23	0.08
AS				1	0.08	0	-0.09	-0.02	-0.02	0.08
MI			1	0.19	0.62	0	-0.13	0.22	0.2	0.11
EPL		1	-0.03	0.04	-0.04	-0.03	0	-0.05	-0.03	0.02
EL	1	0.07	-0.06	-0.01	-0.01	-0.01	0.01	-0.06	-0.05	0.02

Pearson Correlation
-1.0 -0.5 0.0 0.5 1.0

Figure 3. Pearson's r correlation matrix comparing the genetic overlap between the investigated PRS. Note: EL = exceptional longevity [30], EPL = exceptional parental longevity [8], MI = myocardial infarction [31]; AS = all stroke, CAD = coronary artery disease [33], EssHT = essential hypertension [www.nealelab.is], HDL = high-density lipoprotein, LDL = low-density lipoprotein, TC = total cholesterol and TG = total triglyceride [34]. PRS were calculated using a p-value threshold of $p < 5 \times 10^{-5}$.

4. Discussion

In summary, this study did not confirm the hypothesis that ELL individuals have lower polygenic risk scores for cardiovascular-related phenotypes. Only the HDL cholesterol and triglyceride PRS were nominally significantly associated with ELL participants. In contrast and as expected, ELL individuals had higher polygenic risk scores for EL.

In regards to the associations of the various cardiovascular PRS with EL, no findings survived correction for multiple testing. This is despite validating the utility of the lipid PRS by confirming positive associations with measured lipid levels in our sample. Interestingly, the different lipid PRS were based on GWAS that found a large number of genome-wide significant loci (Table 1). ELL individuals had lower LDL and total cholesterol levels than controls in this study, but they did not differ on their respective PRS. This may suggest that environmental factors, perhaps lifestyle-related, influenced these lipid levels, which possibly promote longevity. The most significant finding in our study was for the HDL PRS, with higher scores associated with EL, which was in the expected direction (OR = 1.15, p = 0.034). Other findings that were nominally significant, or approached significance, that were also in the expected direction were the PRS for essential hypertension (OR = 0.89, p = 0.068) and LDL (OR = 0.89, p = 0.077). However, the direction of the relationship between the triglyceride PRS and EL was contrary to expectations (OR = 1.14, p = 0.050). Moreover, the relationships between EL and cardiovascular PRS constructed at less stringent GWAS p-value thresholds reached statistical significance with EL, although not always in the expected direction. In contrast, the UK Biobank study observed that extreme parental longevity (defined as at least one parent who survived to the top 1% of age at death) had lower polygenic risk for several cardiovascular health measures. Namely coronary artery disease, systolic blood pressure, body mass index, high-density lipoproteins, low-density lipoproteins and triglycerides. A similar result for HDL cholesterol and extreme parental longevity (EPL) by the UK Biobank to the current study was reported (OR = 1.08) [8]. Again, similar results were reported by the UK Biobank for LDL (OR = 0.89). However, the observed discrepancies between our analysis and the UK Biobank were most likely due to methodological differences, including the use of PRS that were based on different GWAS and p-value thresholds ($p < 5 \times 10^{-5}$ vs. $p < 5 \times 10^{-8}$, respectively). Additionally, sample sizes varied (e.g., cases: n = 294 vs. n = 1339, respectively) and there were differences in the definitions of EL ($\geq$95 years vs. participants with at least one long-lived parent).

This study confirmed significant associations between higher-longevity PRS and EL in our cohort using summary data from two different GWAS studies, Broer et al. [30] ($\geq$90 years cases) and Pilling et al. [8] (cases had a mother who lived $\geq$98 and a father $\geq$95 years), with the results in the expected direction. Despite both of the longevity PRS not including any genome-wide significant variants ($p < 5 \times 10^{-8}$), PRS scores were significantly associated with EL in our sample. It should be noted that Pilling et al. [8], examining EPL, observed two genome-wide significant hits; however, they were not used in our PRS due to the quality control steps undertaken. Interestingly, 32 SNPs from the EL overlap with 123 SNPs in the EPL at the suggestive threshold of $p < 5 \times 10^{-5}$. EL PRS calculated at other GWAS p-value thresholds were all statistically significant with EL in our sample and showed strengthened associations (e.g., EL: OR increased from 1.2 to 1.8 at a threshold of $p < 5 \times 10^{-5}$ to <0.05). A significant but modest correlation between EL and EPL PRS suggests there is a slight overlap in genetic risk between these two longevity phenotypes. Moreover, when the whole genome ($p \leq 1$) was considered for calculation of the EL and EPL PRS, the correlation coefficient increased to 0.46.

Limitations of this investigation include the relatively small sample size of the current study, which could result in low statistical power. This is demonstrated by a post hoc power analysis based on our observed sample sizes and parameters, which showed that an OR above 1.26 would have provided 80% power, assuming eight independent tests. However, some of the observed ORs were below 1.26 (GWAS p-value threshold $< 5 \times 10^{-5}$), which may have contributed to the observed non-significant results. This study did not examine sex differences, which may influence EL [9]. The other issue is the appropriate GWAS p-value threshold cut-off to use for the analyses. As can be seen from the results described, there is great variation in the results at different thresholds for the same PRS, with some findings in the expected direction, whilst others were contrary. Currently, there is no consensus regarding the use of thresholds, which can have a great influence on the interpretation of the results.

5. Conclusions

Using the current GWAS data available, polygenic risk for cardiovascular-related phenotypes and disease calculated at the 5×10^{-5} threshold appear not to play a strong role in achieving EL. This is despite some evidence that cardiovascular pathways are involved, including lower prevalence of *APOE* ε4 carriers in ELL individuals [9,16]. On the other hand, at less stringent GWAS *p*-value thresholds, there were significant results observed but these were not always in the expected direction. Therefore, studies enrolling larger sample sizes are required to further explore the role of CVD-related genetic variants in EL. Sex and ethnic differences should also be examined. Other EL phenotypes could be investigated, including healthy EL and the use of more extremes of EL (e.g., supercentenarians), which may further reveal the extent of the contribution of cardiovascular genetic determinants to ageing successfully.

Supplementary Materials: The following are available online at http://www.mdpi.com/2073-4425/10/3/227/s1, Table S1: Description of GWAS data sets used for calculation of longevity PRS, Table S2: Description of GWAS data sets used for calculation of cardiovascular PRS, Table S3: The number of SNPs included in calculating a PRS at each *p*-value threshold; Table S4: Logistic regression results for longevity PRS and exceptional longevity in younger controls (55–65 years) vs. long-lived cases (>95 years) at different *p*-value thresholds, Table S5: Logistic regression results for longevity PRS and exceptional longevity in younger controls (55–65 years) vs. long-lived cases (>95 years) at different *p*-value thresholds without the *APOE* locus, Table S6: Logistic regression results for longevity PRS and exceptional longevity in younger controls (55–65 years) vs. long-lived cases (>95 years) at different *p*-value thresholds without the *FOXO3* locus, Table S7: Logistic regression results for longevity PRS and exceptional longevity in younger controls (55–65 years) vs. long-lived cases (>95 years) at different *p*-value thresholds without the *APOC3* locus, Table S8: Logistic regression results for longevity PRS and exceptional longevity in younger controls (55–65 years) vs. long-lived cases (>95 years) at different *p*-value thresholds adjusted for *APOE* ε4 carrier status, Table S9: Logistic regression results for longevity PRS and exceptional longevity in younger controls (55–65 years) vs. long-lived cases (>95 years) at different *p*-value thresholds adjusted for *APOC3*, Table S10: Associations for cardiovascular risk factors (lipids) with measured lipid levels in the combined sample (55–106 years) at various *p*-value thresholds, Table S11: Associations for cardiovascular risk factors (lipids) with measured lipid levels in the combined sample (55–106 years) at various *p*-value thresholds adjusting for *APOE*, Table S12: Associations for cardiovascular risk factors (lipids) with measured lipid levels in the combined sample (55–106 years) at various *p*-value thresholds adjusting for *APOC3*, Table S13: Logistic regression results for the association of cardiovascular PRS and exceptional longevity PRS in younger controls (55–65 years) vs. long-lived cases (>95 years) at different *p*-value thresholds, Table S14: Logistic regression results for the association of cardiovascular PRS and exceptional longevity PRS in younger controls (55–65 years) vs. long-lived cases (>95 years) at different *p*-value thresholds adjusting for *APOE*, Table S15: Logistic regression results for the association of cardiovascular PRS and exceptional longevity PRS in younger controls (55–65 years) vs. long-lived cases (>95 years) at different *p*-value thresholds adjusting for *APOC3* and Table S16: *p*-values for Pearson's correlation matrix comparing the genetic overlap between the investigated PRS (*p*-value threshold of 5×10^{-5}).

Author Contributions: Conceptualisation, K.A.M., A.T., M.R. and P.S.S.; genotyping, T.-J.E., K.A.M., N.J.A., C.O., J.B.K., R.J.S., P.R.S. and J.R.A.; formal analysis, C.O. and A.T.; investigation, M.R.; resources, P.R.S., H.B., P.S.S., J.R.A. and R.J.S.; data curation K.A.M., A.T., N.J.A. and C.R.; writing—original draft preparation, M.R.; writing—review and editing, All; supervision, A.T. and K.A.M.; funding acquisition, K.A.M., P.R.S., H.B., P.S.S., J.R.A. and J.K.

Funding: The Sydney Centenarian Study is supported by the National Health Medical Research Council (NHMRC) Program Grant 109308 and Project Grant 630593. The Hunter Community Study would like to acknowledge the support of the University of Newcastle and the Fairfax Family Foundation. The NHMRC Program Grants 350833, 568969 and 109308, support the Sydney Memory and Ageing Study.

Acknowledgments: We thank and acknowledge the important contributions of the participants and their supporters to the individual studies. We also thank the research teams of the Sydney Centenarian Study, the Sydney Memory and Ageing Study and the Hunter Community Study.

Conflicts of Interest: The authors declare no conflict of interest.

References

1. Oeppen, J.; Vaupel, J.W. Demography. Broken limits to life expectancy. *Science* **2002**, *296*, 1029–1031. [CrossRef]

2. Fries, J.F. Aging, natural death, and the compression of morbidity. 1980. *Bull. World Health Organ* **2002**, *80*, 245–250.
3. Franceschi, C.; Passarino, G.; Mari, D.; Monti, D. Centenarians as a 21st century healthy aging model: A legacy of humanity and the need for a world-wide consortium (WWC100+). *Mech. Ageing Dev.* **2017**, *165*, 55–58. [CrossRef]
4. Atkins, J.L.; Delgado, J.; Pilling, L.C.; Bowman, K.; Masoli, J.A.H.; Kuchel, G.A.; Ferrucci, L.; Melzer, D. Impact of Low Cardiovascular Risk Profiles on Geriatric Outcomes: Evidence From 421,000 Participants in Two Cohorts. *J. Gerontol. A Biol. Sci. Med. Sci.* **2018**. [CrossRef]
5. Vaarhorst, A.A.; Beekman, M.; Suchiman, E.H.; van Heemst, D.; Houwing-Duistermaat, J.J.; Westendorp, R.G.; Slagboom, P.E.; Heijmans, B.T. Leiden Longevity Study Group. Lipid metabolism in long-lived families: The Leiden Longevity Study. *Age* **2011**, *33*, 219–227. [CrossRef] [PubMed]
6. Barzilai, N.; Atzmon, G.; Schechter, C.; Schaefer, E.J.; Cupples, A.L.; Lipton, R.; Cheng, S.; Shuldiner, A.R. Unique lipoprotein phenotype and genotype associated with exceptional longevity. *JAMA* **2003**, *290*, 2030–2040. [CrossRef] [PubMed]
7. Muenchhoff, J.; Song, F.; Poljak, A.; Crawford, J.D.; Mather, K.A.; Kochan, N.A.; Yang, Z.; Trollor, J.N.; Reppermund, S.; Maston, K.; et al. Plasma apolipoproteins and physical and cognitive health in very old individuals. *Neurobiol. Aging* **2017**, *55*, 49–60. [CrossRef]
8. Pilling, L.C.; Atkins, J.L.; Bowman, K.; Jones, S.E.; Tyrrell, J.; Beaumont, R.N.; Ruth, K.S.; Tuke, H.; Yaghootkar, M.A.; Wood, A.R.; et al. Human longevity is influenced by many genetic variants: Evidence from 75,000 UK Biobank participants. *Aging* **2016**, *8*, 547–560. [CrossRef] [PubMed]
9. Revelas, M.; Thalamuthu, A.; Oldmeadow, C.; Evans, T.J.; Armstrong, N.J.; Kwok, J.B.; Brodaty, H.; Schofield, P.R.; Scott, R.J.; Sachdev, P.S.; et al. Review and meta-analysis of genetic polymorphisms associated with exceptional human longevity. *Mech. Ageing Dev.* **2018**, *175*, 24–34. [CrossRef]
10. Dominiczak, M.H.; Caslake, M.J. Apolipoproteins: Metabolic role and clinical biochemistry applications. *Ann. Clin. Biochem.* **2011**, *48*, 498–515. [CrossRef]
11. Lahoz, C.; Schaefer, E.J.; Cupples, L.A.; Wilson, P.W.; Levy, D.; Osgood, D.; Parpos, S.; Pedro-Botet, J.; Daly, J.A.; Ordovas, J.M. Apolipoprotein E genotype and cardiovascular disease in the Framingham Heart Study. *Atherosclerosis* **2001**, *154*, 529–537. [CrossRef]
12. Morris, B.J.; Willcox, D.C.; Donlon, T.A.; Willcox, B.J. FOXO3: A Major Gene for Human Longevity—A Mini-Review. *Gerontology* **2015**, *61*, 515–525. [CrossRef] [PubMed]
13. Willcox, B.J.; Morris, B.J.; Tranah, G.J.; Chen, R.; Masaki, K.H.; He, Q.; Willcox, D.C.; Allsopp, R.C.; Moisyadi, S.; Gerschenson, M.; et al. Longevity-Associated FOXO3 Genotype and its Impact on Coronary Artery Disease Mortality in Japanese, Whites, and Blacks: A Prospective Study of Three American Populations. *J. Gerontol. A Biol. Sci. Med. Sci.* **2017**, *72*, 724–728. [CrossRef] [PubMed]
14. Vergani, C.; Lucchi, T.; Caloni, M.; Ceconi, I.; Calabresi, C.; Scurati, S.; Arosio, B. I405V polymorphism of the cholesteryl ester transfer protein (CETP) gene in young and very old people. *Arch. Gerontol. Geriat.* **2006**, *43*, 213–221. [CrossRef] [PubMed]
15. Atzmon, G.; Rincon, M.; Schechter, C.B.; Shuldiner, A.R.; Lipton, R.B.; Bergman, A.; Barzilai, N. Lipoprotein genotype and conserved pathway for exceptional longevity in humans. *PLoS Biol.* **2006**, *4*, e113. [CrossRef]
16. Panza, F.; D'Introno, A.; Colacicco, A.M.; Capurso, C.; Capurso, S.; Kehoe, P.G.; Capurso, A.; Solfrizzi, V. Vascular genetic factors and human longevity. *Mech. Ageing Dev.* **2004**, *125*, 169–178. [CrossRef]
17. Sachdev, P.S.; Levitan, C.; Crawford, J.; Sidhu, M.; Slavin, M.; Richmond, R.; Kochan, N.; Brodaty, H.; Wen, W.; Kang, K.; et al. The Sydney Centenarian Study: Methodology and profile of centenarians and near-centenarians. *Int. Psychogeriatr.* **2013**, *25*, 993–1005. [CrossRef] [PubMed]
18. Sachdev, P.S.; Brodaty, H.; Reppermund, S.; Kochan, N.A.; Trollor, J.N.; Draper, B.; Slavin, M.J.; Crawford, J.; Kang, K.; Broe, G.A.; et al. The Sydney Memory and Ageing Study (MAS): Methodology and baseline medical and neuropsychiatric characteristics of an elderly epidemiological non-demented cohort of Australians aged 70-90 years. *Int. Psychogeriatr.* **2010**, *22*, 1248–1264. [CrossRef]
19. McEvoy, M.; Smith, W.; D'Este, C.; Duke, J.; Peel, R.; Schofield, P.; Scott, R.; Byles, J.; Henry, D.; Ewald, B.; et al. Cohort profile: The Hunter Community Study. *Int. J. Epidemiol.* **2010**, *39*, 1452–1463. [CrossRef]
20. Song, F.; Poljak, A.; Crawford, J.; Kochan, N.A.; Wen, W.; Cameron, B.; Lux, O.; Brodaty, H.; Mather, K.; Smythe, G.A.; et al. Plasma apolipoprotein levels are associated with cognitive status and decline in a community cohort of older individuals. *PLoS ONE* **2012**, *7*, e34078. [CrossRef] [PubMed]

21. Muenchhoff, J.; Poljak, A.; Song, F.; Raftery, M.; Brodaty, H.; Duncan, M.; McEvoy, M.; Attia, J.; Schofield, P.W.; Sachdev, P.S. Plasma protein profiling of mild cognitive impairment and Alzheimer's disease across two independent cohorts. *J. Alzheimers Dis.* **2015**, *43*, 1355–1373. [CrossRef]

22. Price, A.L.; Patterson, N.J.; Plenge, R.M.; Weinblatt, M.E.; Shadick, N.A.; Reich, D. Principal components analysis corrects for stratification in genome-wide association studies. *Nat. Genet.* **2006**, *38*, 904–909. [CrossRef]

23. Das, S.; Forer, L.; Schonherr, S.; Sidore, C.; Locke, A.E.; Kwong, A.; Vrieze, S.I.; Chew, E.Y.; Levy, S.; McGue, M.; et al. Next-generation genotype imputation service and methods. *Nat. Genet.* **2016**, *48*, 1284–1287. [CrossRef]

24. Oldmeadow, C.; Holliday, E.G.; McEvoy, M.; Scott, R.; Kwok, J.B.; Mather, K.; Sachdev, P.; Schofield, P.; Attia, J. Concordance between direct and imputed *APOE* genotypes using 1000 Genomes data. *J. Alzheimers Dis.* **2014**, *42*, 391–393. [CrossRef] [PubMed]

25. Euesden, J.; Lewis, C.M.; O'Reilly, P.F. PRSice: Polygenic Risk Score software. *Bioinformatics* **2015**, *31*, 1466–1468. [CrossRef] [PubMed]

26. R Core Team. *R: A Language and Environment for Statistical Computing*; R Core Team: Vienna, Austria, 2017.

27. Nagelkerke, N.J.D. A Note on a General Definition of the Coefficient of Determination. *Biometrika* **1991**, *78*, 691–692. [CrossRef]

28. Zhang, D. *rsq: R-Squared and Related Measures*, R Package version 1.1; Vienna, Austria, 2018.

29. Zhang, M. *WebPower: Basic and Advanced Statistical Power Analysis*, R Package version 0.5.2; Vienna, Austria, 2018.

30. Broer, L.; Buchman, A.S.; Deelen, J.; Evans, D.S.; Faul, J.D.; Lunetta, K.L.; Sebastiani, P.; Smith, J.A.; Smith, A.V.; Tanaka, T.; et al. GWAS of longevity in CHARGE consortium confirms APOE and FOXO3 candidacy. *J. Gerontol. A Biol. Sci. Med. Sci.* **2015**, *70*, 110–118. [CrossRef]

31. Myocardial Infarction, G.; Investigators, C.A.E.C.; Stitziel, N.O.; Stirrups, K.E.; Masca, N.G.; Erdmann, J.; Ferrario, P.G.; Konig, I.R.; Weeke, P.E.; Webb, T.R.; et al. Coding Variation in ANGPTL4, LPL, and SVEP1 and the Risk of Coronary Disease. *N. Engl. J. Med.* **2016**, *374*, 1134–1144. [CrossRef]

32. Malik, R.; Chauhan, G.; Traylor, M.; Sargurupremraj, M.; Okada, Y.; Mishra, A.; Rutten-Jacobs, L.; Giese, A.K.; van der Laan, S.W.; Gretarsdottir, S.; et al. Multiancestry genome-wide association study of 520,000 subjects identifies 32 loci associated with stroke and stroke subtypes. *Nat. Genet.* **2018**, *50*, 524–537. [CrossRef] [PubMed]

33. Nelson, C.P.; Goel, A.; Butterworth, A.S.; Kanoni, S.; Webb, T.R.; Marouli, E.; Zeng, L.; Ntalla, I.; Lai, F.Y.; Hopewell, J.C.; et al. Association analyses based on false discovery rate implicate new loci for coronary artery disease. *Nat. Genet.* **2017**, *49*, 1385–1391. [CrossRef] [PubMed]

34. Willer, C.J.; Schmidt, E.M.; Sengupta, S.; Peloso, G.M.; Gustafsson, S.; Kanoni, S.; Ganna, A.; Chen, J.; Buchkovich, M.L.; Mora, S.; et al. Discovery and refinement of loci associated with lipid levels. *Nat. Genet.* **2013**, *45*, 1274–1283. [CrossRef] [PubMed]

MDPI

St. Alban-Anlage 66

4052 Basel

Switzerland

Tel. +41 61 683 77 34

Fax +41 61 302 89 18

www.mdpi.com

Genes Editorial Office

E-mail: genes@mdpi.com

www.mdpi.com/journal/genes

www.ingramcontent.com/pod-product-compliance
Lightning Source LLC
Chambersburg PA
CBHW051913210326
41597CB00033B/6131